ANALYTICAL TECHNIQUES IN MEAT SCIENCE

Analytical Techniques in Meat Science is a comprehensive compilation of all the relevant methodologies for the quality analysis of meat. The content of the book is designed to cater to requirement of meat producers, regulatory agencies, researchers, students, teachers, laboratory staff etc. It covers techniques for physico-chemical analysis, species identification and microbiological examination of meat. Also, it contains the latest biotechnological and proteomic techniques for meat quality evaluation. To help the reader understand better figures, tables, line diagrams, etc are used frequently whenever needed. Some important pictures are given in plates for lucid and clear understanding of the concept.

Dr. Vivek Vinayak Kulkarni is a Veterinarian and Ph.D in Livestock Products Technology (Meat Science). He has 30 years of professional experience and has worked in six states in various capacities. Dr.Kulkarni recently retired as Professor and Head of LTP dept in Veterinary College and Research Institute, Namakkal (TANUVAS).

Dr. Girish Patil, S., M.V.Sc., Ph.D., MBA is working as Principal Scientist at ICAR – National Research Centre on Meat, Hyderabad. The frontier areas in which he initiated research for the first time in India are meat species identification, meat traceability, application of information technology for safe meat production, organic meat production and lab grown meat production.

Dr. Sukhadeo B Barbuddhe, M.V.Sc., Ph.D., Director, ICAR-National Research Centre on Meat, Hyderabad has completed B.V.Sc. and A.H. from Nagpur Veterinary College, Nagpur in 1991. Subsequently, he joined Indian Veterinary Research Institute, Izatnagar for higher studies and completed M.V.Sc. and Ph.D. in 1993 and 1996, respectively. He has been working on *Listeria monocytogenes* since last two decades.

Dr. Naveena B. Maheswarappa, M.V.Sc., Ph.D. is working as Principal Scientist at ICAR – National Research Centre on Meat, Hyderabad. He has established the proteomics laboratory at ICAR-NRC on Meat and contributed significantly in the area of high-throughput proteomic tools for understanding muscle food quality, lipid-protein interaction and identified the protein biomarkers for meat colour and texture.

Dr. M. Muthukumar, M.V.Sc., Ph.D. is working as Principal Scientist at ICAR - National Research Centre on Meat, Hyderabad. He has 2 years of experience in integrated meat processing plants and 15 years of research, extension and teaching experience in meat science and technology. He has worked on 9 extramural projects sponsored by MoSPI, MoFPI, DBT, DADH, FSSAI, APEDA and ICAR and 22 institute projects.

ANALYTICAL TECHNIQUES IN MEAT SCIENCE

V. V. Kulkarni
P. S. Girish
S. B. Barbuddhe
B. M. Naveena
M. Muthukumar

NARENDRA PUBLISHING HOUSE
DELHI (INDIA)

First published 2022
by CRC Press
2 Park Square, Milton Park, Abingdon, Oxon, OX14 4RN

and by CRC Press
6000 Broken Sound Parkway NW, Suite 300, Boca Raton, FL 33487-2742

© 2022 Narendra Publishing House

CRC Press is an imprint of Informa UK Limited

The right of V. V. Kulkarni et.al. to be identified as authors of this work has been asserted by him in accordance with sections 77 and 78 of the Copyright, Designs and Patents Act 1988.

Print edition not for sale in South Asia (India, Sri Lanka, Nepal, Bangladesh, Pakistan or Bhutan).

British Library Cataloguing-in-Publication Data
A catalogue record for this book is available from the British Library

Library of Congress Cataloging-in-Publication Data
A catalog record has been requested

ISBN: 978-1-032-13801-5 (hbk)
ISBN: 978-1-003-23089-2 (ebk)

DOI: 10.1201/9781003230892

MIX
Paper from
responsible sources
FSC
www.fsc.org   FSC® C013985

Printed in the United Kingdom
by Henry Ling Limited

Contents

Preface

Meat is an important component of food basket of India and it plays important role in nutritional and nutrient security of the country. Demand for meat and meat products is rapidly increasing in the domestic market due to rising per capita income, ever growing population and increasing nuclear families. It is pertinent to note that quality consciousness among domestic consumers is slowly rising due to increasing education and awareness level. Stringent food safety guidelines putforth by Food Safety & Standards Authority of India (FSSAI) has made it mandatorty for meat producers and processors to maintain quality and safety standards of meat and meat products. Also, in recent years India is emerging as a major player in international meat market with buffalo meat export reaching record high. Meat producers need to maintain quality and ensure safety to comply with domestic regulations, meeting consumer expectations in Indian scenario and to meet the regulations of the importing countries in the International market. This requires techniques for evaluation of meat quality and availability of laboratory facilities in different parts of the country to meet the analytical requirements of the growing meat industry.

Analytical Techniques in Meat Science is a comprehensive compilation of all the relevant methodologies for the quality analysis of meat. The content of the book is designed to cater to requirement of meat producers, regulatory agencies, researchers, students, teachers, laboratory staff etc. The book is the largest compilation of the techniques in meat quality evaluation in the country. It covers techniques for physico-chemical analysis, species identification and microbiological examination of meat. Also, it contains the latest biotechnological and proteomic techniques for meat quality evaluation. To help the reader understand better figures, tables, line diagrams, etc are used frequently whenever needed. Some important pictures are given in plates for lucid and clear understanding of the concept. Appendices are included at the end of the text which helps to understand the text better. We hope that our humble effort should prove successful in helping the readers. We have tried to make the topics as comprehensive and simplified as possible. However, certain inaccuracies and errors may creep in for which suggestions and comments will be appreciated greatly.

Authors

C H A P T E R - 1

■ ■ ■

SAMPLING OF MEAT FOR PHYSICO-CHEMICAL ANALYSIS

1.1 SAMPLING PROCEDURE

Sampling is a process for making an inference from analytical data, through multiple mass reduction stages from a decision unit/sampling target. Material from which sample is collected and inference made is termed as decision unit. Individual portion of material collected by a single operation of a sampling tool and combined with other increments to form a primary sample constitutes the increment.

The collection of one or more increments taken from a decision unit according to a sampling protocol is the primary sample and when it is received in laboratory for analysis, it becomes laboratory sample. From this analytical sample and test portions can be drawn.

A sample is a small part taken from a larger amount of material for examination. The purpose of sampling meat is to obtain representative material on which the tests are to be carried out for composition and other specifications to compare with those required e.g. microbiological counts, residue contents, etc. On the basis of the tests the samples can be accepted or rejected. It is, therefore, very important that the samples taken are truly representative of the whole batch; otherwise good material may be rejected and poor materials accepted.

In the analysis of any biological material, the most important step is the preparation of a representative and uniform sample. Highly accurate analytical results are valueless if obtained from a poor sample.

Random Sampling

This method is normally used in the sampling of meat supplies. In this, a limited number of units from the batch are sampled at random. It is important that the selection of the units to be sampled is truly random and that the selection is not biased e.g. ease of access, unusual appearance or habits. In the absence of specific instructions about number of samples, the square root of the number of items in the batch is often taken as the number of samples required.

Acceptable Sampling

These plans are commonly used in the sampling of raw materials and finished products for acceptance. Acceptance sampling provides information on whether a particular lot of materials meet predetermined acceptance sampling plan to provide the buyer with measured assurance of obtaining raw material or finished goods of a desired quality and a probability of rejection of inferior material. These plans are based on the statistical design and structured for specific product.

Sample Labeling

Whenever samples are taken they must be labeled. The label should be legible, indelible and securely attached to the sample container or package. The information required on the label should include, either in writing or in code.

a) time and date of sampling

b) identity of the supplier

c) batch no.

d) material description

e) sample description (taken from surface, core or dip, etc.)

1.2 SAMPLE DOCUMENTATION AND HANDLING

In any sampling and testing operation there are at least three activities, usually separated in location and time: 1. Sampling 2. Testing and evaluation of results 3. Communication of results for decisions and action.

There needs to be an organised system of communication between sampling and laboratory procedures. Often these functions will be carried out by different personnel. To prevent unnecessary delays and possible sample deterioration, the laboratory needs to know in advance the number of samples and the precise tests required. Careful consideration must be given to sample transport and storage.

Sample should not be exposed to direct sunlight and should not be transported at elevated temperatures.

It is obviously impossible to describe in detail sampling procedure required for all the different meat products, under many different conditions, prevalent in the meat packing industry.

Specific technique must be developed by each laboratory to fit to their products and conditions.

1.3 SAMPLING EQUIPMENT

Sampling tools and equipment are themselves potential source of foreign bodies in meat. Hence they should be kept as a kit in a separate box and should be checked before and after the sampling. Sampling tools should be sufficiently robust that it does not break easily causing hazard to the staff. All reusable tools e.g. knives, drills, spatulas etc. must be easy to clean and such that it does not contaminate the sample.

Laboratory Items Required

1. Meat grinder with set of plates and knives
2. Waring Blender with set of several speeds
3. Refrigerator
4. Knives, drills and meat saw
5. Cups and lids of polystyrene
6. Glass jars with screw covers
7. Spatulas of various sizes for mixing samples

1.4 PRECAUTIONS WHILE COLLECTING THE SAMPLES

1. To avoid serious error due to loss of moisture during the preparation and subsequent handling, do not attempt to prepare small samples.
2. Immediately after taking a sample, place it in a glass jar with tight fitting cover.
3. Do not allow the sample to come in contact with paper of any kind, especially easily absorbing moisture.
4. Label the sample completely for identification.
5. Keep the sample refrigerated at all times even at the time of preparation and in transit.

6. It is essential that condensation be avoided, which may occur when sample is moved from a refrigerator to a warm area.

For Unground Samples

1. Remove any bones present from the sample before grinding, such as ham.
2. Grind the samples in a cooler if possible. Put the sample through a meat grinder two or three times, mixing thoroughly each time. Use a plate with approximately 1 / 8 inch holes for all grindings, if possible. If the sample is difficult to grind use the coarser plate for the first grinding.
3. Include the residue left in the chopper and mix well after each grinding.
4. Grind and mix the sample rapidly and transfer to an airtight glass container or cup with tight lid and be sure that no air is left inside the container for possible accumulation of moisture from sample.
5. Store in refrigerator and analyse as soon as possible.

Frozen meat

Sampling of frozen meat presents a greater problem than fresh meat. To inspect the interior of frozen meat block, it is cut with a band saw into nine equal cubes, these cubes provide a total of 54 cut surfaces for visual examination. When it is necessary to remove samples of material from frozen block, cores of meat can be removed using angular drills. Five hole of 2.5 cm diameter and at least $3/4^{th}$ cm depth of the block are removed in the Domino five patterns.

References

Church, P.N. and Wood, J.M. (1992) In the manual of manufacturing meat quality, Elsevier Applied Science Publishers Limited, Crown house, Linton road, Barking, Essex, IG11 8JU, England.

Kowale, B.N., Kulkarni, V.V. and Kesava Rao, V. (2008) Methods in meat science. *Pub.* by Jaypee brothers medical publishers (P) Ltd., Ansari Road, Daryagunj, New Delhi.

■ ■ ■
MOISTURE ESTIMATION IN MEAT

2.1 SAMPLE PREPARATION

Separate fat, fascia and connective tissue from meat and mince it through 3-mm plate, mix thoroughly. This sample can be used for various estimations promptly or it can be preserved in polyethylene packs at low temperature depending upon the period for which it is used. Sample must be minced, mixed, macerated or ground so that they are homogenous. Meat samples must be weighed immediately after mincing because water may separate out within a few hours.

2.2 HOT AIR OVEN METHOD

Introduction

Water constitutes about 65-80 percent of total muscle. It is present in the space between the myofilaments and held by capillary forces in the form of free (15 percent) and bound water. The free water is mainly responsible for the organoleptic qualities of meat while bound water is generally unaffected by various stages of processing of meat. The water-soluble sarcoplasmic proteins, vitamins, myoglobin, glycogen, and other enzymes are suspended in muscle water.

Principle

The sample is dried at 100- 105°C for 16 to18 h and the loss of weight is reported as moisture. In this procedure, volatile substances present in the sample are also accounted as water.

Equipment

Hot air oven, constant weight moisture cups with lid, fine analytical balance, and desiccator.

Procedure

Weigh 10 g of the meat sample into a weighed moisture cup and dry in a preheated oven at 100 to 105°C for 16 to18 hours and cool in a desiccator. The process of heating and cooling is repeated till a constant weight is achieved.

Calculation

$$\text{Moisture percent} = \frac{\text{Initial weight -- Final weight}}{\text{Weight of the sample}} \times 100$$

2.3 RAPID MICROWAVE DRYING METHOD

Principle

Moisture is removed (evaporated) from sample by using microwave energy. Weight loss is determined by electronic balance readings before and after drying and is converted to moisture content by microprocessor with digital percent readout.

Apparatus

a) Microwave moisture analyser - 0.2 mg H_2O sensitivity, moisture / solid range of 0.1 to 99.9 percent, 0.01 percent resolution. Includes automatic tare electronic balance, microwave drying system and microprocessor digital computer control. Electronic balance pan is located inside drying chamber. (balance sensitivity: - 0.2 mg at 15 g capacity or 1.0 mg at 40 g capacity.

b) Glass fibre pads - 9.8 x 10.2 cm rectangular glass fibre pads or equivalent.

Procedure

Prepare sample and place 2 rectangular glass fibre pads on balance pan in microwave moisture analyser drying chamber and tare. Remove pads from chamber, rapidly and evenly deposit 4 g well mixed sample on rough side of one pad. Place second pad over sample and replace pads and sample on balance pan. Dry sample with pad 3-5 min at 80 to 100 percent power, depending on product type. On completion of microwave drying cycle, read percent moisture displayed on digital readout panel.

Certain product classes require addition of adjustment factor to readout for accurate results e.g. cooked sausages, pre-blends / emulsions, cured / cooked meats, factor = 0.55 percent.

Reference

AOAC (1995).

■ ■ ■

MEAT PROTEIN ANALYSIS

3.1 CRUDE PROTEIN ESTIMATION

3.1.1 Micro - Kjeldahl Method

Principle

Meat sample is digested in H_2SO_4 using $CuSO_4.5H_2O$ as catalyst. Na_2SO_4 (anhydrous) or K_2SO_4 elevate the boiling point and convert organically bound N to ammonium sulphate which when heated with excess alkali (40percent NaOH), NH_3 is liberated which is distilled into known excess of standard acid. $(0.1NH_2SO_4)$. The unreacted (un-neutralised) acid is back titrated with standard alkali (0.1N,NaOH). The titre values give the nitrogen content from which crude protein is calculated by multiplying with a factor of 6.25.

$$2\ NH_3 + H_2SO_4 \longrightarrow (NH_4)_2SO_4$$

$$(NH_4)_2\ SO_4 + 2\ NaOH \longrightarrow 2NH_3 + Na_2SO_4 + 2H_2O$$

Apparatus

Kjeldahl flask, balance, digestion chamber, Micro Kjeldahl's distillation assembly (or Kjeltec), volumetric flask, burette, pipette and beaker / conical flask.

Reagents

1. Digestion mixture (9.5 part sodium sulphate and 0.5 parts copper sulphate).
2. Nitrogen free concentrated H_2SO_4
3. 40 percent NaOH (40g in 100 ml of solution)

4. N / 10 H_2SO_4 (2.78 ml in one liter of distilled water and standardised with 0.1 N NaOH using Methyl red as indicator.

5. Toshiro's indicator: Dissolve 80 mg of methyl red and 20 mg of bromocresol green in 100 ml alcohol. Add 10 ml of this solution to one liter of 0.2 percent boric acid solution and mix.

Procedure

1. Take 2 g of sample in Kjeldahl flask and add a pinch of digestion mixture and add 20 ml of nitrogen free concentrated sulphuric acid.

2. Digest till the solution turns green / straw yellow colour. Cool the flask.

3. Carefully transfer the contents into a 250 ml volumetric flask. Wash repeatedly with NH_3 free distilled water and collect the washings also into the volumetric flask. Make the volume up to the mark and cool under tap.

4. Take 10 ml of the sample aliquot, in a Micro Kjeldahl distillation unit.

5. Add about 20 ml of 40 percent NaOH. Take 10 ml of Toshiro's indicator in a conical flask and dip the tip of silver tube into it. Heat the alkaline liquid by passing steam into it and 30 ml of distillate is collected in the conical flask. Remove the conical flask after rinsing the tip with a little distilled water.

6. Titrate the distillate with standard N/10 H_2SO_4 to a light pink end point. Always perform a blank test and correct the reading accordingly.

7. From the titration values, the percentage of protein is calculated.

Calculation

Percent protein (crude) = X × 0.0014 × 250 × 100 × 6.25 / weight of sample × 10 × 1 (AOAC, 1995).

NOTE

Reagent proportions, heat input, and digestion time are critical factors and do not change.

Copper sulphate is added to speed up the conversion of organically bound nitrogen (protein, amino acid, amines, amides) into ammonium sulphate.

Potassium sulphate or anhydrous sodium sulphate is added to elevate the boiling point, which hastens the digestion, and avoid loss of H_2SO_4 due to evaporation.

3.1.2 Biuret Method

Principle

The - CONH groups in the protein molecule react with copper sulphate in alkaline medium to give purple colour which is then read at 540 nm.

Reagents

1. Biuret Reagent: Weigh 6 g of Potassium tartrate ($KC_2H_4O_6$, $4H_2O$), 1.5 g of cupric sulphate ($CuSO_4$ $5H_2O$) and with constant stirring add 300 ml of 10 percent NaOH and dilute to 1000 ml with distilled water.
2. Standard: The standard protein solution may be either a pooled normal human serum (standardised by Kjeldahl method) or a solution of pure albumin in saline (1mg /ml).

Procedure

To 1 ml aliquots of standard, test sample and blank (saline or distilled water) add 5 ml of Biuret reagent. Mix well and keep for 30 minutes. Read absorbance of test and standard against blank at 540 nm.

Calculation

Total protein g / 100 ml = OD of the sample × Concentration of standard / OD of standard (Gornall *et al.*, 1949).

3.1.3 Lowry's Procedure

Principle

The final colour is a result of, (a) Biuret reaction of protein with copper ion in alkali and (b) reduction of the phosphotungstic acid by the tyrosine and tryptophan present in the treated protein.

Reagents

1. **3N Folin-Ciocalteau reagent**: Weigh 100 g sodium tungstate and 25 g of sodium molybdate and transfer in a 1000 ml volumetric flask. Add 700 ml of distilled water to dissolve the chemicals and add 50 ml of 85 percent Orthophosphoric acid (specific gravity 1.75) followed by 100 ml HCl. Reflux

the solution for 10 hours and cool and add 150 g of Lithium sulphate, distilled water 50 ml and few drops of bromine water. Boil the solution without condenser for 15 min, cool and make the volume up to 1000 ml and store in a dark bottle. This stock solution should be diluted with two volumes of distilled water before use.

2. 4 percent Na_2CO_3

3. 0.5 percent $CuSO_4.5H_2O$ in 1percent potassium sodium tartrate

4. Alkaline copper solution. Mix 50 ml of 4 percent Na_2CO_3 (reagent 2) with 2 ml of 0.5percent $CuSO_4.5H_2O$ in 1percent potassium sodium tartrate (reagent 3). Prepare fresh.

5. 0.1 N NaOH

6. Diluted Folins' Reagent. Dilute Folin-Ciocalteau reagent with an equal volume of 0.1 N NaOH.

7. Standard protein solution: The standard solution is prepared with bovine serum albumin to have a concentration of 1mg protein/ml.

Procedure

Take aliquot or protein solution as 100 mg, 200 mg, 300 mg in different test tubes and add distilled water to make volume 1 ml. Pipette 1 ml of the test solution and 1 ml of 0.1N NaOH for blank. Add 5 ml of freshly prepared cooper sulphate reagent. Incubate for 10 min, add 0.5 ml of diluted Folin-Ciocalteau reagent and incubate for 30 min. Read the colour (bluish green) at 670 nm.

The unknown sample is then read from the standard curve and the necessary calculations made (Lowey *et al.,* 1951).

3.2 FRACTIONATION OF MUSCLE PROTEINS ON THE BASIS OF SOLUBILITY

Introduction

Muscle proteins based on their solubility can be classified as water-soluble proteins (sarcoplasmic), salt soluble proteins (myofibrillar) and acid soluble proteins (collagen, elastin, reticulin), with a percent proportion of 5 percent. 11.5 percent and 2.5 percent, respectively, out of total 19 percent. The myofibrillar proteins play a vital role in meat processing as an emulsifier of fat and binder. During the storage of meat, amongst the proteins the water-soluble proteins are attacked by the proteolytic enzymes of bacteria while the structural proteins are the last to be proteolysed. The percent protein and the break down products are used as an indicator of extent of proteolysis during storage of meat.

Principle

The protein fractions are extracted in the suitable medium depending upon the maximum extractability and quantitation is done by estimating nitrogen by micro Kjeldahl method.

Equipment

Digestion flask, micro Kjeldahl apparatus, spectrophotometer and refrigerated centrifuge.

Procedure

Sarcoplasmic proteins: - Weigh 20 g meat, add 30 ml of chilled distilled water and blend in a homogeniser for 2 min. Transfer the homogenate quantitatively to volumetric flask and make the volume to 200 ml. Keep overnight at 4 to 5°C and centrifuge at 3000 rpm for 15 min. in refrigerated centrifuge. Measure nitrogen in the supernatant by microkjeldahl method.

Myofibrillar proteins: - Dispense the residue from above estimation in 200 ml of 0.67 M NaCl solution. Keep overnight at 4 to 5°C and centrifuge at 3000 rpm for 15 min. in refrigerated centrifuge. Measure nitrogen in the supernatant by microkjeldahl method.

Stroma proteins: - Estimate nitrogen in the residue by microkjeldahl method (Kang and Pice, 1970).

3.3 ESTIMATION OF STROMA PROTEINS

The residue obtained after extraction of myofibrillar proteins (Kang and Rice, 1970) can be further processed for extraction of stroma proteins according to Bendall (1967) and Narayanan and Page (1976).

The residue is defatted by stirring in a large excess of acetone for 24 hours at 4°C. The residue is finally washed in ether and centrifuged. The defatted residue is homogenised with 10 ml of 0.5 M acetic acid and kept overnight at 0 to 4°C. The supernatant is collected by centrifugation and labelled as acid extractable stroma proteins.

The residue is further extracted by stirring in 10 ml of 0.1 N sodium hydroxide at 95 to 98°C for 1 hour and left overnight at 4°C. It is centrifuged next day and supernatant is collected and labelled as alkali extractable stroma protein. The insoluble stroma protein fraction (residue) is dried with acetone and ether. Weighed and expressed as g per 100 g wet tissue (Prasad *et al.*, 1990; Bendau, 1967 and Narayan and Pape, 1976).

3.4 COLLAGEN ESTIMATION

The major component of connective tissue is collagen, which contains the amino acid L-hydroxyproline in much higher levels than found in the muscles.

Hydroxyproline is a derived amino acid present only in connective tissue and highest in collagen (12 to 18 percent). It is, therefore, used as a measure of the connective tissue. It is formed by incorporation of ascorbic acid into the proline polypeptide chain. Hydroxyproline concentration in beef muscles is reported to range between 200 and 770 µg / g. Although excessive proteolysis of the collagen and elastin of connective tissue might appear to be the most likely change causing increased tenderness, the portion of connective tissue is not normally changed in this way during conditioning in skeletal muscle. This was evidenced by no increase in water-soluble hydroxyproline, even after storage of sterile, fresh meat for one year at 37°C. By measuring L –hydroxyproline content the connective tissue content of raw meat can be estimated. Hydroxyproline content of meat is determined according to the procedure of Stegeman and Stadler (1967) modified by International organisation for standardisation (1974). The collagen content is calculated by multiplying hydroxyproline content with 7.14.

3.4.1 Hydroxyproline Estimation by Chloramin T Method

Principle

Meat is hydrolysed in a solution of hydrochloric acid. The resultant hydroxyproline is oxidised with chloramin –T and reacted with 4- dimethyl aminobenzaldehyde (DMEB). The red colour developed is measured photometrically.

Reagents

1. HCL, 6N
2. NaOH, 10N solution (40g of NaOH dissolve in water and dilute to 100 ml of distilled water).
3. Poranol-1
4. Iso-proply alcohol (peropanol –2)
5. Buffer solution, pH 6.0

Dissolve in water 50 g of citric acid monohydrate, 12 ml acetic acid, 120 g of sodium acetate trihydrate and 34 g of sodium hydroxide and dilute to 1 liter with water.

Oxidation Reagent

Dissolve 1.41 g of Chloramin T, in 10 ml of water, 10 ml of propanol-1 and 80 ml of buffer and keep it in a dark bottle at 4°C.

Colour Reagent

Dissolve 10g, p-dimethyl aminobenzadehyde (DMEB) in 35 ml of perchloric acid solution (60 percent) and then slowly add 65 ml of Propanol –2. Prepare this solution on the day of use.

Procedure

Preparation of hydrolysate: - Transfer accurately weighed 4 g of thoroughly mixed muscle sample to a boiling flask (200ml) attached with a water-cooled condenser. Add 100 ml 6N HCl and continue hydrolysis on a water bath for 6 to 8 hours. Filter the hot hydrolysate through a sintered glass funnel (200 ml G-4), wash the flask 3 times with 6N HCl and add to filtrate as above. Dilute the hydrolysate up to 200 ml. Take 5 ml of the hydrolysate in a glass beaker and nutralise with 10 percent NaOH initially followed by 1N NaOH adjusting pH to 6.8. Make up the volume of the neutralised hydrolysate to 100 ml. Take 5 ml of aliquot (hydrolysate) in a test tube and add 2 ml of oxidation reagent. Mix the contents on a shaker and keep at room temperature for 20 minutes. Add 2 ml of the freshly prepared colour reagent and mix. Transfer the test tube to water bath (maintained at 60°C) for exactly 15 minutes. Cool the test tubes under running water for at least 3 minutes and read the absorbance at 560 nm immediately in a spectrophotometer. The concentration can be calculated from the regression equation computed from the reference values.

Reference Values for Hydroxyproline

Weigh accurately 100 mg of standard hydroxyproline in a 100 ml volumetric flask. Add little water and a drop of 6N HCl. Make the volume with distilled water. Keep in refrigerator and use as stock solution. Make 5 serial dilutions in a test tube using 0.5, 1.0, 1.5, 2.0 and 2.5 mg / ml and follow the dilutions as described above adding oxidising reagent and colouring agents. Read the colour at 560 nm for calculating regression equation.

Calculation

$$\text{Hydroxyproline (mg/100 g meat)} = \frac{\text{O.D. of unknown} \times \text{Conc. of standard}}{\text{O.D. of standard}} \times 100$$

(Nueman and Logan, 1950)

Collagen percent (wet weight) = 7.25 × Hydroxyproline (mg / 100 g)

From the calibration graph the calculations is as given below

Percent hydroxyproline = 5C / MV

C = concentration of diluted hydrolysate in mg / ml.

M = weight of sample in g.

V = volume of aliquot of hydrolysate taken for dilution in ml.

Collagen percent (wet weight) = 7.25 × hydroxyproline (mg /100 g)

3.4.2 Estimation of collagen content

Hydroxyproline (HP) content of the meat sample was determined based on the procedure of Nueman & Logan (1950) with few modifications as suggested by Naveena & Mendiratta (2001). Two-gram meat sample was hydrolysed with 40 ml of 6 N, HCl for 18 h at 108°C. The hydrolysate was filtered and the volume adjusted to 50 ml with distilled water. Then 25 ml of hydrolysate was taken and pH was adjusted to 7.0 using 40% NaOH and the volume was adjusted to 50 ml again with distilled water. One ml of aliquot from this solution was used for hydroxyproline estimation. Absorbance was measured at 540 nm using UV-VIS spectrophotometer (Model: UV-1700 PharmaSpec, SHIMADZU, Japan) and the hydroxyproline content was determined by referring to a standard graph. Collagen content was calculated by multiplying hydroxyproline content with 7.14 and was expressed in mg/g tissue.

3.4.3 Estimation of collagen solubility

Collagen solubility was determined by the method described by Mahendrakar, Dani, Ramesh, & Amla (1989). Five grams of muscle tissue was taken in a 250ml beaker and immersed in water bath after covering the beaker with watch-glass. The water bath was then heated to boiling temperature and held for 30 min. The cooked meat was then taken out of the beaker and cut into small pieces and homogenized with 50 ml distilled water at 4±1°C in a blender for 2 min. The extract was then centrifuged at 4000 rpm for 30 min. Aliquots of cooked out juice and centrifugate were hydrolyzed for 18 h at 108°C in hot air oven and soluble HP was calculated according to Williams & Harrison (1978).

% HP Solubilized = (g HP in drip + g HP in cooked meat/g HP in raw meat) × 100

% Collagen solubility = 7.14 × % HP solubilised

2g meat + 40ml 6N HCl
↓
Keep it in hot air oven for 18 hrs at 105°C
↓
Filter and adjust the volume to 50ml with distilled water
↓
Take 25ml aliquot and adjust the P^H 7 with 40% NaOH
↓
Adjust volume to 50ml with distilled water
↓
Take 1ml

	SAMPLE	BLANK
Sample ⟶	1ml	1ml D.W
0.01M CuSO$_4$ ⟶	1ml	1ml
2.5N NaOH ⟶	1ml	1ml
6% H$_2$O$_2$ ⟶	1ml	1ml

Mix and keep at room temperature for 5min with occasional shaking
↓
Place tubes at 80°C in water bath for 5min with occasional shaking
↓
Chilled on ice
↓
Add 4ml of 3N sulfuric acid (H$_2$SO$_4$) in each tube
↓
Add 2ml of 5% DMEB in n-propanol in each tube
↓
Mix well and place all the tubes in water bath at 70°C for 16min
↓
Cool and take OD at 540nm.
↓
HP content expressed in % mg/g of tissue by standard graph

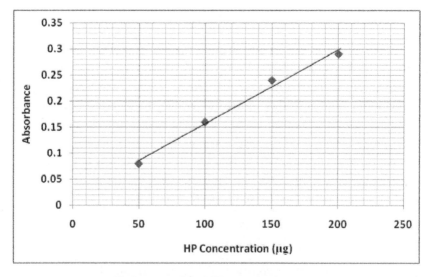

Hydroxyproline (HP) standard curve

Collagen solubility was calculated as described below (Dransfield *et al.*, 1983).

5g raw/cooked meat in 250ml beaker

↓

Immerse it in water bath at 100°C for 30min

↓

Sample cut into small pieces and add 30ml distilled water

↓

Centrifuge at 4000rpm for 30min

↓

Take Supernatant and add 30ml 6N Hcl

↓

Digestion for 18hrs at 105°C and filter

↓

Take 1ml of filtrate

↓

Follow the procedure similar to collagen content estimation

Calculations of collagen content and solubility

- Meat used for acid hydrolysis ———— 2g
- 2g meat acid hydrolysate is made to 50 ml
- From that 25 ml solution is used
- 25 ml solution now contains 1g meat
- This 25 ml made into 50 ml
- So now 50 ml solution contain 1g meat
- From this 50 ml, only 1ml aliquot is used for hydroxyproline estimation
- Therefore we are calculating HP for 0.02g meat
- Absorbance at 540 nm provides HP content in micro gram (10^{-6})
- Let HP content be A
- 0.02g meat————————————A $\times$ 10^{-6}g of HP
- 1g meat————————————1/2A $\times$ 10^{-4}g of HP
- 100g meat————————————1/2 A $\times$ 10^{-4} $\times$ 10^{2} g of HP
- Converting this into collagen by using 7.25 factor
- 100g meat————————7.25/2 $\times$ A $\times$ 10^{-2} g of collagen
- 3.625×10^{-2} g of collagen

Similar derivation in cooked meat give rise to $2.41 \times A \times 10^{-2}$ g collagen

Result in both cases is in g%

3.4.4 Estimation of protein extractability

Protein extractability was determined according to procedure of Joo *et al.* (1999). Sarcoplasmic and total proteins (sarcoplasmic + myofibrillar) were extracted separately by homogenizing 2 g sample with 20 ml of ice-cold 0.025 M potassium phosphate buffer (pH 7.2) or 20 ml of ice-cold 1.1 M potassium iodide in 0.1M phosphate buffer (pH 7.2) respectively and kept overnight at 4°C with frequent shaking. All Samples were centrifuged at 5000 X *g* for 20 min and concentration of protein in the supernatant was determined by the Biuret method. Myofibrillar protein extractability was calculated by obtaining difference between total and sarcoplasmic protein extractability.

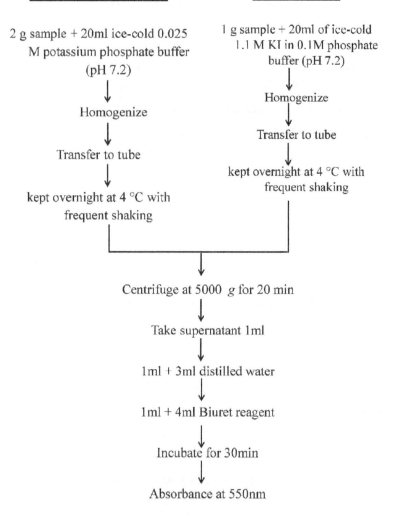

Protein Estimation by Biuret Method

The reaction is based on the action of diluted copper sulfate solution with peptides especially nitrogen in alkaline solution. A purple coloured solution is formed due to the reaction between copper and nitrogen atoms of peptides.

Bovine Serum Albumin (BSA) standard curve preparation

The 200 mg of BSA was dissolved in 10 ml distilled water to get 20 mg protein per ml. From this stock solution 0.1, 0.2, 0.4, 0.6, 0.8 and 1 ml solution were transferred to dry test tube and 0.9, 0.8, 0.6, 0.4, 0.2 and 0 ml of distilled water was added. The volume was made to 1 ml in each tube to contain a protein concentration of 2, 4, 8, 12, 16 and 20 mg/ml respectively. Add 4 ml Biuret reagent to each tube and mix well. After colour development the absorbance was determined using UV-VIS spectrophotometer (Model: UV-1700 PharmaSpec, SHIMADZU, Japan) at 550 nm and plotted against protein concentration on a graph.

Calculation of protein extractability

Sarcoplasmic proteins: If 1ml supernatant gives an absorbance of 0.22 at 550 nm, the proteins concentration is roughly 5 mg/ml by looking at the aforementioned standard graph.

i.e 1ml supernatant contains 5 mg protein

Total of 20ml (or 2 gm meat) contains 100mg protein

That means 1 g meat contains 50 mg proteins

1 g meat is equivalent to 0.2 g protein (considering approx. 20% protein)

i.e. 0.2 g protein contains 50 mg extractable proteins

or 250 mg sarcoplasmic protein is extractable from 1.0 g protein

Total proteins

i.e 1ml supernatant contains 5 mg protein

Total of 20ml (or 1 gm meat) contains 100 mg protein

That means 1 g meat contains 100 mg proteins

1 g meat is equivalent to 0.2 g protein (considering approx. 20% protein)

i.e. 0.2 g protein contains 100 mg extractable proteins

or 500 mg total protein is extractable from 1.0 g protein

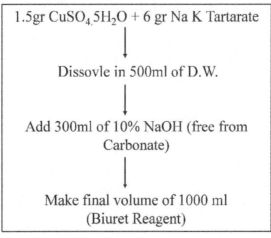

1.5gr CuSO$_4$.5H$_2$O + 6 gr Na K Tartarate

↓

Dissovle in 500ml of D.W.

↓

Add 300ml of 10% NaOH (free from Carbonate)

↓

Make final volume of 1000 ml (Biuret Reagent)

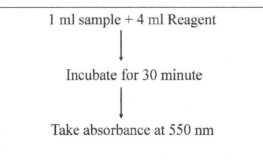

1 ml sample + 4 ml Reagent

↓

Incubate for 30 minute

↓

Take absorbance at 550 nm

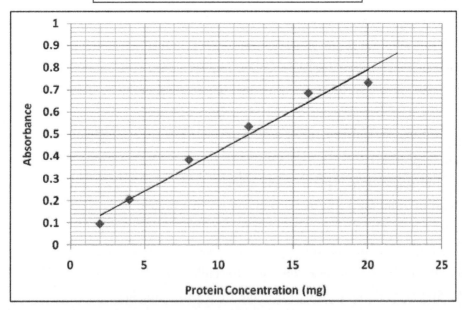

Bovine serum albumin (BSA) standard curve

References

AOAC (1995).

Bendall, J.R. (1967) J.Fd. Agric., 18, 553-558.

Dransfield, E., Cassy J.C., Boccard R., Touraille, C. (1983). Comparison of chemical composition of meat determined at eight laboratories, Meat Science 8(2): 79-192.

Gornall, A.G.; Bradwill, C.J. and David, M.M. (1949) J. Biol. Chem., 177: 751.

International organisation of standardisation (1974) Draft international standard ISO / DIS- 3496

Joo, S. T., Kauffman, R. G., Kim, B. C., and Park, G. B. (1999). The relationship of sarcoplasmic and myofibrillar protein solubility to colour and water holding capacity in porcine longissimus muscle. Meat Sci.35,276-278

Kang, C.K. and Rice, E.E. Journal of Food Science (1970) 35- 563-565.

Lowry, O.H., Rosenbrough, N.J., Farr, A.L. and Randall, P.J., (1951) J.Biol. Chem., 196: 265-27.

Mahendrakar, N.S. Dani, N.P., Ramesh, B.S. and Amla, B.L. (1989). Studies on influence of age of sheep postmortem carcass conditioning treatment, on muscular cohagen content and it, thermostability. Journal of Food Science & Technol. 26: 102-106.

Narayanan, A.S. and Page, R.C. (1976), Biol. Chem., 251- 1125-1130.

Naveena, B.M. and Mendiratta, S.K. (2001). Tenderization of Spenthen meat using gives extract. British Poultry Science, 42: 344-349.

Naveena, B.M., Kiran, M., Sudakar Reddy, K. Vaithiyanathan, S., Ramakrishna, C. and Suresh K. Devatkal. (2011). Effect of ammonium hydroxide on ultrastructure and tenderness of buffalo meat. *Meat Science,* 88, 727-732.

Naveena, B.M., Kiran, M., Sudakar Reddy, K. Vaithiyanathan, S., Ramakrishna, C. and Suresh K. Devatkal. (2011). Effect of ammonium hydroxide on ultrastructure and tenderness of buffalo meat. *Meat Science,* 88, 727-732.

Nueman, R. E., & Logan, M.A. (1950). Determination of Hydroxyproline content. *Journal of Biological Chemistry,* 184, 299-306.

Prasad, R.L. and Singh, L.N. (1990) Ind. J. Meat Sci. and Technol., 3 (1) 22-28.

Stegman, H and Stadler, K (1967) Clin. Chem. Acta., 18: 167

Williams, J. R., & Harrison, D. L. (1978). Relationship of hydroxyproline solubility to tenderness of bovine muscles. Journal of Food Science, 43, 864.

Woessner, J.F. (1961) Arch. Biochem. Biophysics., 93: 440-447.

C H A P T E R - 4
■ ■ ■
MEAT LIPID ANALYSIS

Lipid in muscle is mainly present in the inter and intramuscular spaces. Fat is responsible for the flavour and taste of meat and is also considered as an important constituent for assessment of keeping quality of meat at various temperatures through development of various oxidation products of fat and fatty acids leading to rancidity. Fat content of meat varies from 3 to 9 percent for lean tissue. The fat of adipose tissue consists almost entirely of true fat or triglyceride. The main fatty acids present in the meat fat are oleic, palmitic, stearic and linoleic. Apart from the triglycerides, animal fat contains small proportions of phospholipids, sterols, carotenoid pigments and fat-soluble vitamins. In general, males have less intramuscular fat than females and castrated members of each sex have more intramuscular fat than corresponding sexually entire animals. In high plane of nutrition, percentage of intramuscular fat increases as more fat is synthesised from carbohydrates and such fat has consequently low iodine number, whereas increasing degree of emaciation, the relative percentage of linoleic acid increases and that of palmitic acid decreases.

Phospholipids are the integral part of cellular membranes and exist in tissues as phospholipoproteins. Phospholipid content depends upon the total lipid content. With the increase in total lipids from 3 to10 percent a decrease in phospholipid of 5 to10 percent has been reported. However, phospholipid fraction is less variable than the total lipid content. Of the total phospholipid in beef muscles, lecithin, accounts for about 62 percent, cephalin 30 percent and sphingomyelin less than 10percent.

Phospholipid content of bovine muscles represents about 28 percent of total lipids from unfrozen muscles and approximately one percent on wet tissue basis. Although the content of phospholipid in beef muscle is quite low, they are more susceptible to oxidation and this makes them important in governing the quality i.e. off flavour, off odour and discolouration.

4.1 DETERMINATION OF ETHER EXTRACT / CRUDE FAT

Principle

The crude fat (a combination of simple fat, fatty acids, ester compound fat, neutral fat, sterols, waxes, vitamins (A, D2, E, K), carotene, chlorophyll etc.) is estimated by extracting in non-aqueous solvents (petroleum ether) continuously in a Soxhlet's apparatus.

Apparatus

Soxhlet's apparatus, thimble with cotton swab, hot air oven, hot plate, balance, weight box and desiccator.

Solvent

Petroleum ether solvent (B.P. 60 - 80°C)

Procedure

1. Set the Soxhlet's apparatus. Weigh the round bottom (R.B.) flask.
2. Weigh 3-4 g sample into fat extraction thimble containing small amount of sand or asbestos. Mix with glass rod, place thimble and rod in 50 ml beaker and dry in oven for 6 hours at 100°C to 102°C or 1.5 hours 125°C. Plug the open end of the thimble with absorbent cotton.
3. Fix the thimble, the open end up into the extractor of the set.
4. Connect the water condenser and circulate cold water through the condenser.
5. Pour petroleum ether (60 - 80°C boiling range) from the top of the condenser so as to have sufficient excess in the R.B. flask after filling the extraction chamber.
6. Heat the flask over a hot water bath (or heating mantles available for the purpose). The extraction is allowed to continue for four hours at a condensation rate of 5 to 6 drops per second.
7. The Round bottom flask is separated after recovery of the solvent and dried in hot air oven at 100 + 5°C for sometime, cooled in a desiccator and weighed.
8. The increase in weight of the R.B. flask gives the crude fat content, which is expressed in percentage.
9. Alternatively the thimble is cooled to room temperature after keeping in hot air oven at 100 - 105°C and weighed. By knowing the difference in weight of the sample the crude fat is calculated.

Calculation

I.   Percentage of crude fat = gain in weight of the round bottom flask / weight of the sample × 100

II. Percentage of crude fat = weight of fat / weight of sample × 100

Note:

1. To prevent ether loss, put cotton swab on the mouth of the condenser.

2. The thimble should be placed above the siphon tube in Soxhlet apparatus.

3. The ether obtained by distillation can be reused (AOAC, 1975).

4.2 LIPID EXTRACTION

Weigh 2 g of meat accurately and transfer quantitatively into Erlenmeyer flask with 20 volumes of solvent mixture comprising of Chloroform: Methanol, (2:1,V/V). Homogenize in a high-speed homogeniser for 1 min and allow to stand at room temperature with occasional stirring for 6 to 8 hours. Filter the extract through Whatman No.1 filter paper and re-extract the residue with 10 more volumes of the same solvent mixture for 2 hr and filter. Combine the two filtrates and evaporate to dryness in vacuum at 45°C in a rotary evaporator. For breaking the proteolipids, dissolve the dried residue in one-tenth volume of solvent mixture containing Chloroform: Methanol: Water (64:32:4, v/v/v) and evaporate to dryness as above. Repeat this step twice and dissolve the residue in 50 ml of Chloroform: Methanol (2:1,V/V) and filter in separating funnel. Add 10 ml of 0.9percent NaCl to remove non-lipid impurities and allow to stand overnight. Collect the Chloroform layer and evaporate to dryness as above and make to a known volume in Chloroform (5ml). To each lipid solution in a glass stoppered tube, add a drop of 0.5 percent Butylated Hydroxy Toluene (BHT) in Chloroform. Store the samples at - 20°C.

4.3 ESTIMATION OF TOTAL LIPIDS (GRAVIMETRICALLY)

Pipette 1 ml of lipid sample into stainless steel planchet (with constant weight predetermined) and dry under an infrared lamp till nearly dry. Then dry at 60°C in an oven to a constant weight. Subtraction of initial and final weight gives the weight of total lipid and express as mg/g of tissue (Folch and Sloane-Stanley, (1957).

4.4 ESTIMATION OF TOTAL PHOSPHOLIPIDS BY ANSA METHOD

Principle

Phospholipids are quantitated by determination of the phosphorus content of the lipid extract by converting into inorganic phosphorus. Most procedures for estimation of phosphorus are based on the method of Fiske and Subba Row (1925), which involves conversion of inorganic phosphate to phosphomolybdate and its subsequent reduction to molybdenum blue.

Reagents and solutions

ANSA: Dissolve 0.5 g of aminonapthol sulphonic acid in 200 ml of 15 percent anhydrous sodium bisulphite followed by addition of 1 g of anhydrous sodium sulphite. This reagent is stable for one week when stored at 4°C in brown bottle.

Ammonium molybdate Solution

Dissolve 25 g of ammonium molybdate in 200 ml of water, add 300 ml of 10N H_2SO_4. Make up to 1 liter.

Standard phosphorus solution

Dissolve 0.351 g of KH_2PO_4 in 10 ml of 10N H_2SO_4 make up to 1 liter. This contains 0.4 mg of phosphorus / 5 ml.

Procedure

Pipette accurately 25 ml of lipid sample (prepared from 2 g of meat and lipid sample volume made to 5 ml with chloroform), 50 m l standard phosphate (KH_2PO_4) solution containing 4 mg phosphorus separately in test tube and evaporate to dryness, add 1 ml of 60percent perchloric acid to each test tube and digest over sand bath for 15 min. For blank use 1 ml of 60 percent perchloric acid. Allow the test tubes to cool at room temperature. To each tube add 0.5 ml of 2.5percent ammonium molybdate followed by 0.2 ml of ANSA and 7 ml of distilled water. Keep in boiling water bath for 7 min, cool and read the blue colour developed at 830 nm. Phospholipid content can be calculated by multiplying the inorganic phosphorus content with the factor of 25 and expressed as mg per g of tissue.

Calculation

$$\frac{\text{OD of unknown} \times \text{Conc. of std.} \times \text{Volume of lipid sample} \times 1}{\text{of standard} \times \text{Volume used for estimation} \times \text{wt of sample} \times 1000}$$

= mg of phosphorus / g, mg of phospholipid / g = 25 x mg P / g

(Marinetti, 1962)

4.5 ESTIMATION OF CHOLESTROL IN MEAT

Cholesterol is a minor but important component of animal tissue occurring in free form or esterified in combination with fatty acids. The concentration of total cholesterol in bovine muscle is reported to range between 58 and 67 mg / 100 g of muscle and is resistant to deterioration during frozen storage. Cholesterol concentration in offal (particularly brain) is much more (1000-1200 mg / 100g) than that in muscular tissue.

4.5.1 Estimation of Total Cholesterol by Acetyl Chloride Method

Equipment and Reagents

- Spectrophotometer, water bath
- $ZnCl_2$ Reagent - dissolve 40g of anhydrous $ZnCl_2$ in 150 ml glacial acetic acid at 80°C for 2.5 hours Filter through glass wool
- Standard cholesterol: 1 mg / ml in Chloroform

Procedure

Pipette 100 ml of lipid extract (prepared from 2 g of meat and lipid sample volume made to 5 ml with chloroform) and 50 ml of standard cholesterol separately in test tube and evaporate to dryness. Add 2 ml of Chloroform, 1 ml of $ZnCl_2$ reagent and 1 ml acetyl chloride. Heat in water bath at 50°C for 10 min. For blank use 2 ml Chloroform, 1 ml each $ZnCl_2$ and acetyl chloride. Read optical density (pink red colour) at 528 nm in a spectrophotometer.

Calculation

$$\text{mg Cholesterol} / 100\text{g of muscle tissue} = \frac{\text{O.D. of unknown} \times \text{Conc. of standard} \times 5}{\text{O.D. of standard} \times 0.1 \times 2 \times 1000}$$

Normal value :- 50 - 65 mg / 100 g muscle tissue (Hanel and Dam, 1955).

4.5.2 Estimation Total Cholesterol by $FeCl_3$- H_2SO_4 Method

Solutions

- Standard Cholesterol solution (1 mg /ml in distilled ethanol)
- $FeCl_3$ stock solution:- 10 g $FeCl_3$ in 100 ml acetic acid

- FeCl$_3$-H$_2$SO$_4$ reagent: - 2 ml of FeCl$_3$ solution diluted to 200 ml with concentrated H$_2$SO$_4$.

- 33 percent KOH (W/W): 10 g of KOH dissolved in water and volume made to 30 ml.

- Alcoholic KOH:- 6 ml of 33 percent KOH made up to 100 ml with distilled ethanol. This solution is prepared fresh before use.

Take 200 ml (0.2 ml) of lipid extract (prepared from 2 g of meat and lipid sample volume made to 5 ml with chloroform) and evaporate to dryness. Add 5 ml of alcoholic KOH and incubate at 37°C for 55 minutes, cool to room temperature. Add 10 ml petroleum ether and invert tube to mix the contents. Add 5 ml distilled water, shake vigorously for 1 min. Take 2 ml aliquot from supernatant (petroleum ether) and evaporate petroleum ether. For standard (1 mg /ml) take 0.1 ml (100 ml) and evaporate. Add glacial acetic acid 3 ml and 2 ml FeCl$_3$ reagent. Mix, keep for 1/2 hour and read at 570 nm. Cholesterol content is expressed as mg of Cholesterol / 100 g of meat. For blank take 30 ml acetic acid.

Calculation

$$\text{mg Cholesterol} /100 \text{ g muscle tissue} = \frac{\text{OD of unknown} \times \text{Conc. of standard} \times 5 \times 100}{\text{OD of standard} \times 0.2 \times 2 \times 10 \times 1000}$$

(Raghuramulu *et al.*, 1983)

4.6 FRACTIONATION OF LIPIDS INTO VARIOUS CLASSES BY TLC

The choice of method for separation of lipid classes into their individual members depends on the desired degree of separation, amount of material available, number of samples and the time factor involved. Several schemes involving, complex formation, solvent fractionation counter current distribution, salt fractionation, column chromatography, thin layer chromatography and impregnated paper chromatography have been employed to fractionate lipid mixtures (Hanahan, 1960; Rouser *et al.*, 1961; Marinetti, 1962; Stahl, 1965; Misra, 1966 and Wagner, 1967).

To prepare a silica gel G coated plates, weigh 30 g of silica gel G in a glass stoppered Erlenmeyer flask. Add 63 ml 0.01 N Sodium carbonate solution. Shake vigorously for 90 seconds. Pour the slurry into the spreader preset at 250 microns thickness and coat the plates. Allow setting for 10 minutes and activate the plates by heating at 110°C in hot air oven for 90 minutes and cool in desiccator chamber. Divide the plates into 3 cm wide lanes and draw horizontal line 17 cm from one end of the plate. Apply sample into streak 2 cm from unmarked end.

4.6.1 Separation of Neutral Lipids

Apply aliquots of lipid extract containing approximately 200 µg of cholesterol on TLC plates along with standards viz., free and esterified cholesterol, tristearin, dipalmitin, monopalmitin and palmitic acid. Develop the plates in a uni-dimensional triple development system using n-hexane: diethyl either: acetic acid. The solvent systems shown in Table 4.1 can be used (Misra, 1968). Develop the plates first for 7.5 cm from the origin in the solvent 'A' and then air dry. Then develop for 15 cm in the solvent system 'B'. The solvent system 'C' can be used optionally to run the plate up to 1.5 cm from the origin, whenever the separation of mono-glycerides from the phospholipids is incomplete. A typical neutral lipid separation has been shown in plate 1.

Table 4.1: Solvent system used in TLC for separation of neutral lipids and phospholipids.

Solvent	Solvent constituent	Solvent ratio	Used for
A	Hexane : Diethyl ether: Glacial acetic acid	60:40:1	Developing solvent for neutral lipid
B	Hexane: Diethyl ether: Glacial acetic acid	90:10:1	Developing solvent for neutral lipid
C	Hexane: Diethyl ether: Glacial acetic acid	30:70:1	Developing solvent for neutral lipid
D	Hexane: Diethyl ether	1:1	Eluting neutral lipids from gel
E	Chloroform	1	Eluting cholesterol and cholesterol fraction from gel
F	Chloroform: Methanol: 7 M NH_4OH	115:45:7.5	For separation of individual phospholipids

Misra U.K. (1968) J.Sci. Ind. Res.25: 303- 318.

Detection of Lipid Spots

Locate the lipid spots on the chromatoplates by exposing the dry chromatoplates to iodine vapours in closed chamber. The neutral lipid spots can be identified by comparing their R_f values with authentic standards co-chromatographed with each run.

Elution of Neutral Lipid Spots and Their Analysis

After exposing the chromatoplates to iodine vapours, mark the spots with needle and allow the iodine to evaporate. Scrape the gel containing the lipid spot carefully with the help of a knife onto a glazed paper and transfer to 20 ml test tubes. For eluting free and esterified cholesterol, add 10 ml of chloroform to the corresponding tubes and for eluting mono-di and tri-glycerides, add 5 to 10 ml of hexane: ether (1:1, V / V) solvent mixture to corresponding tubes. Stir the contents with a glass rod intermittently for 15 minutes and centrifuge at 2000 rpm for 5 minutes. Decant the supernatant carefully into another set of tubes and repeat the extraction thrice. Evaporate the combined supernatants to dryness at 60_0C and dissolve the residues in a known volume of chloroform. Extract gel blanks for each lipid spot similarly and make appropriate corrections in the corresponding values of lipid components. Analysis of cholesterol can be done by the method of Hanel and Dam (1955) and glyceride glycerol by the method of Van Handel and Zilversmit (1957).

4.6.2 Separation of Glycerides

Glycerides are quantitated by number of methods based on gravimetry, colourimetry and chromatographic analysis. Total glycerides are also calculated indirectly by subtracting the sum of total phospholipids and total cholesterol from total lipid values. Though in this indirect method non-esterified fatty acids, hydrocarbons and fat soluble substances contribute to the value of glycerides, still it is meaningful for expressing glycerides, because the concentrations of these substances in the tissue is quite small. Other methods of glyceride estimation are based upon chromatographic isolation of individual glycerides and the chemical estimation of glyceride-glycerol. After alkaline hydrolysis, the glyceride-glycerol is oxidised with periodate to formaldehyde. The formaldehyde is measured colourimetrically after reaction with chromotropic acid or phenyl hydrazine (Wagner, 1967). The liberated glycerol can also be condensed with 'O' aminophenol, which in turn is oxidised to yield 8-hydroxy quinoline, which is, measured fluorimetrically.

Glycerides are better assayed by glyceride-glycerol estimation, according to Van Handel and Zilversmit (1957).

Reagents

1.   0.1N alcoholic potassium hydroxide,
2.   0.4 N Sulphuric acid,
3.   66 percent Sulphuric acid,
4.   0.05 M sodium periodate,
5.   20 percent sodium sulphite,

6. Chromatropic acid (0.5 g chromatropic acid dissolved in distilled water, after completely dissolving add 250 ml 66 percent sulphuric acid.

7. Standard tryglyceride (weigh 0.2 mg tristerin and dissolve in 100 ml chloroform).

Procedure

Take the lipid sample in a test tube and evaporate to dryness in vacuum. Then add 0.5 ml of 0.1N alcoholic potassium hydroxide and mix well and keep at 70°C for 20 min to facilitate saponification. After saponification remove the tubes and add 0.2 ml of 0.4 N H_2SO_4 to neutralise excess of KOH. Now add 0.1 ml of 0.05 M sodium periodate to oxidise the fatty acids to aldehyde. Wait for 10 min, so that all fatty acids are oxidised. Then add 0.2 ml of 20 percent sodium sulphite solution and mix well. Again wait for 10 min and add 8 ml of chromotropic acid reagent. Mix well and keep in boiling water bath for 30 min. Cool the tubes. Read the absorbance at 570 nm and express glyceride glycerol as mg/g of tissue. Appropriate standard of tristearin is carried through the same procedure along with a blank. For blank take 0.1 ml chloroform and process the same way as mentioned above.

Total Glycerides

Total glycerides can be indirectly calculated by subtracting the sum of total phospholipids, total cholesterol, total glycolipids and total free fatty acids from total lipid values.

4.7 FRACTIONATION OF PHOSPHOLIPIDS BY THIN LAYER CHROMATOGRAPHY

Prepare silica gel G coated plates as mentioned above. Activate the plates by heating at 110°C in hot air oven for 90 minutes and cool in desiccator chamber. Divide the plates into 3 cm wide lanes and draw horizontal lines 17 cm from one end of the plate. Apply sample into streak 2 cm from unmarked end.

Apply aliquots of lipid extract containing approximately 30 micrograms of phospholipid phosphorus on TLC plates along with standard containing Lysophosphatidyl choline (LPC), Lysophosphatidyl ethanolamine (LPE), Sphingomyelin (SPH) Phosphatidyl choline (PC) and Phosphatidyl ethanolamine (PE). For developing the plates use solvent system comprising of Chloroform: Methanol: 7 M Ammonium Hydroxide (230:90:15 v / v / v) (Abramson and Blecher, 1964) Remove the plates when the solvent mixture reaches the 15 cm line from the origin and air dry. For localisation and identification of phospholipid spots on the plates expose them to iodine vapour in closed chamber. The identification can be done by comparing R_f values with authentic phospholipid standard co-chromatographed with each run. Scrap carefully the gel containing phospholipid spot and transfer quantitatively in a 25 ml test tube separately. Direct gel digestion method of Misra, (1969) may be followed for estimation of phospholipid phosphorus of various fractions. In this method, the gel containing phospholipid is digested with 1 ml of 60 percent perchloric acid in a 25 ml test tube on hot sand bath for 20 minutes. The colour is developed according to the modified method of Marinetti, (1962). The gel can be sedimented by centrifugation at 2000 rpm for 2 minutes after colour development. The phospholipid fractions are expressed as percent of total phospholipids (Misra, 1969).

4.8 FATTY ACID ANALYSIS

Separation of Neutral Lipids and Phosphlipids by column Chromatography: (Separation of neutral lipids (NL) and phospholipids (PL) can be carried out by adopting the method of Hornstein *et al.*, (1961) using silicic acid column chromatography with slight modification.). Activate ten g of silicic acid (100 to 200 mesh) overnight at 130°C and cool in a desiccator. To this add chloroform and pour into a 2.5 x 90 cm glass column fitted with a sintered glass disc. Remove air bubbles by stirring the mixture with a long glass rod. Allow the silicic acid to settle and drain the chloroform. Wash the column with 100 ml chloroform.

Apply the lipid extract containing approximately 250 to 300 mg of total lipids over the column. As soon as lipid migrated inside the column, place a little cotton. Elute the NL with 10 bed volumes of solvent containing hexane: diethyl ether (1:1, V/V) in two strokes. Elute the Phospholipids (PL) with 10 bed volumes of methanol in two strokes. Evaporate both the fractions to dryness under vacuum in a rotary evaporator at 45°C.

Preparation of Methyl esters

Add 30 ml of alcoholic KOH (10 per cent) to the dried neutral lipid and phospholipids extracts and allow to stay in the dark overnight. Next day, saponify samples further under reflux for about one hour. Cool the contents of the flask and transfer into separating funnel. Add thirty ml of distilled water and 50 ml of hexane. After thorough mixing, allow to separate the two phases. Collect the lower phase in the same flask and discard the upper phase, which contained non-saponifiables. To the lower phase in the flasks add 15 ml HCl and 50 ml hexane, mix and allow separating in a separating funnel. Collect Hexane (the upper layer) in a separate flask. The lower phase is collected and again washed with 30 ml distilled ware and 50 ml hexane, mix and allow to separate. Collect the hexane layer in the hexane flask and discard lower phase. Evaporate the collected Hexane to dryness in the rotary evaporator.

Add 50 ml of 3 percent methanolic HCl to the vacuum dried fatty acid samples and reflux for 1 hour. Add few glass beads to avoid bumping. Remove flasks and transfer the contents to a separating funnel, in which 30 ml of distilled water and 50 ml hexane are added and mix and allow to separate. Collect the lower phase into the same flask and collect the hexane layer in a separate flask. Repeat the procedure twice and evaporate the collected hexane and dry up to 2 to 3 ml. Final concentrations of methyl esters in hexane shall be 1 mg/ml. The samples of methyl esters shall be stored at -18°C.

Analysis of Fatty Acids from NI, PI and Total Lipids by Gas Liquid Chromatography (GLC)

Fatty acid methyl esters can be analysed isothermally using modular gas chromatograph equipped with a flame ionization detector and a recorder. A coiled stainless steel column (2m×4mm) packed with 20 percent diethyleneglycol adipate (DEGA) on chromosorb W (60-80 mesh) can be used. The operating conditions shall be: column temperature 185°C, injector temperature 220°C, and detector temperature 240°C. Ultra high purity nitrogen is used as carrier gas at a flow rate of 30 ml per minute.

The fatty acid peaks can be identified by comparing the retention time of standard fatty acids on semilog paper. Peak areas can be calculated by triangulation. The peak area corresponding to the fatty acid is expressed as percentage of total peak area. The following fatty acids can be identified in buffalo meat lipids on GLC; capric (C10), lauric (C12), myristoleic (C14:1), palmitic (C16), palmitoleic (C16:1), stearic (C18), oleic (C18:1), linoleic (C18:2), linolenic (18:3) and arachidonic (C20:4). A typical gas liquid chromatogram of methyl esters of neutral lipid and phospholipid fatty acids of buffalo meat is shown in plate1.

4.9 SEPARATION OF CHOLESTEROL AND CHOLESTEROL OXIDES (OXYSTEROLS) BY PREPARATORY THIN LAYER CHROMATOGRAPHY (TLC)

Saponify the dried lipid extract (0.12 to 0.15 g of lipid) with 20 ml of 1N methanolic KOH for 12 hours at room temperature with occasional shaking. Extract the unsaponifiable fraction twice with 25 ml of diethyl ether after adding

20 ml of double distilled water. Rinse the pooled ether extract at least three times with double distilled water until the pH of the aqueous phase is neutral. Evaporate the unsaponifiable fraction to dryness under vacuum and re-suspend into 0.5 ml of chloroform. Technique of thin layer chromatography with silica gel G can be used for separation of cholesterol oxides from cholesterol.

Prepare silica gel coated plates as mentioned above.

Separation of Cholesterol and Cholesterol Oxides

Apply the un-saponifiable fraction on TLC plate along with cholesterol standard. Separate the cholesterol and cholesterol oxides by developing the plates unidimensionally in a solvent system comprising of hexane: Diethyl ether (70:30, V/V). After the solvent had reached the 15 cm line from the origin, remove the plates from chromatographic tank and air dry.

Detection of cholesterol and Cholesterol Oxides

Locate cholesterol and cholesterol oxide by exposing the dry chromatoplates to iodine vapours in closed chamber. Identify the cholesterol oxidation products by comparing R_f value with authentic standard co-chromatographed.

Elution of Cholesterol Oxides (Oxysterols)

After exposing the chromatoplates to iodine vapours, mark the oxysterol spots with needle and allow iodine to evaporate. Scrape the gel containing the oxysterols carefully with the help of knife onto a glazed paper and transfer to 20-ml test tubes. For eluting oxysterols add 10 ml of diethyl ether and stir with a glass rod intermittently for 15 min and centrifuge at 2000 rpm for 5 min. Decant the supernatant carefully into another set of tubes and repeat this extraction process thrice.

Evaporate the combined supernatants to dryness at 40°C and dissolve the oxysterol residues in a known volume of chloroform.

Separation of Cholesterol Oxides (Oxysterols)

Apply the dissolved oxysterols in chloroform on TLC plates along with standards of cholesterol oxides. Separate the oxysterols by developing the plates in a solvent system comprising of hexane: diethyl ether:ethyl acetate (50:50:50, v/v/v) (Pie *et al.*, 1991). After the solvent had reached the 15 cm line from the origin, remove the plates from chromatographic tank and air dry. The following oxysterols can be identified and separated from the origin viz., cholesterol, 7-hydroxycholesterol, 19-hydroxycholesterol, 7-ketocholesterol, cholesterol-epoxide, cholesterol ß-epoxide and unidentified fraction. A typical oxysterol separation is shown in plate-1.

The localization and identification of oxysterols on the chromatoplates can be done by exposure to iodine vapours in a sealed chamber. The identification of oxysterols can be done by comparing the R_f values with authentic oxysterol standards co-chromatographed with each run.

Scrape the gel containing oxysterol spots carefully with the help of knife onto glazed paper and transfer quantitatively into 20 ml test tubes separately. For eluting oxysterols, add 5 ml of 7 percent propanol in hexane to each test tube. Stir the contents with glass rod intermittently for 15 min and centrifuge at 2000 rpm for 5 min. Decant the supernatant carefully into a set of tubes and repeat the extraction thrice. Evaporate the combined supernatants to dryness at 40°C under vacuum.

Quantitative analysis of cholesterol oxides can be done by using a method of Willemart and Ferdet (1959). Add two ml of 0.2 N sodium hydroxide (NaOH) to the dried oxysterol residue followed by 10 ml of phosphoric acid H_3PO_4) and heat at 40°C for 90 min. Make volume 25 ml by adding Butanol and read the optical density at 260 nm and express the oxysterols as per cent of total oxysterols.

4.10 GAS CHROMATOGRAPHIC ANALYSIS OF FATTY ACIDS IN FOODS OF ANIMAL ORIGIN

Gas chromatography is the most widely used analytical technique for estimating the pesticide residues, fatty acid profile, etc. It offers both qualitative and quantitative analysis with speed, accuracy, reproducibility and sensitivity. GC is a basically a partition chromatography where mobile phase is gas and stationary phase is liquid. The separation of compounds occurs due to difference in partition coefficient in vapour form between stationary liquid and mobile gas phases. The basis for gas chromatographic separation is the distribution of a sample between two phases. One of these phase is a stationary phase of large surface and other phase is mobile phase i.e., a gas, which passed through the stationary phase. GC is operated at high temperature (100 – 400°C) and therefore it is used for analysis of compounds that forms varpous and are stable at temperature.

A known quantity of solution (1– 5 ul) is injected into gas chromatographic column through an inlet (Inject port) heated to a sufficient temperature that vaporises the sample solution. In this state, the flow of carrier gas (inert gas) that form a mobile phase sweeps analyte through the column. Nitrogen and Helium gases are most commonly used as carrier gases. During flowing through the column, analyte that were injected in the same solutions separate from one other because of the selective interaction with liquid phase and also properties of the analytes. Proper selection of a column, column oven temperature and the rate of gas flow play important role in the separation of the components present in the injected sample and completing analysis in a reasonable length of time. Sometimes temperature programme is desirable when a mixture of containing analytes with variable boiling points. The packed columns of different materials like glass, stainless steel, aluminum etc are available. The liquid phase is coated on an inert solid material called support in packed columns. In capillary and mega bore column liquid phase is coated on the walls of the fused silica columns. When analyte elute from the column enters into the detector port, where detector responds to the presence of specific element or functional group within the molecule. The detectors response causes a change in electrode potential (signal), which is proportional to the amount of analayte present in the injected solution. The signal is amplified by electrometer and send to recorder where the recorder, record the chromatogram. Electro capture Detector (ECD), Nitrogen Phosphrous Detector (NPD/TID), Flame photometric detector (FPD), Thermal conductivity Detector (TCD) and Flame ionization Detector (FID)are some of the important detectors.

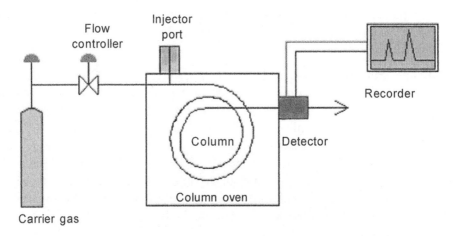

Block diagram of gas chromatography

Determination of Fatty Acids in foods of animal origin

Instruments & Apparatus: 50 ml-Polypropylene centrifuge tubes, vials, conical flasks, reagent bottles, 12 ml test tubes, rotary vacuum evaporator/ nitrogen evaporator, water bath, Centrifuge, Vortexer, PVDF syringe filter (0.22μ), Gas Chromatography –FID.GC Column –SP^{TM} -2560 capillary column-100 x 0.25 mm x $0.2\mu m$, hamilton syringe - 5/10μl.

Chemical/reagents: Supelco 37 component FAME Mix in DCM, 10NKOH, water, methanol, 24N H_2SO_4, anhydrous Na_2SO_4, n-hexane.

Preparation of reagents:

Preparation of10N KOH: 561gms of KOH pellets are to be dissolved and made up to 1000ml distilled water.

Preparation of 24N H_2SO_4: 638.4 ml of H_2SO_4 made upto 1000 ml.

Extraction Procedure: 0.5gms of meat sample is to be taken in 50 ml graduated conical bottomed centrifuge tubes. Add 700 μl of 10N KOH and 5.3ml of Methanol to these tubes. Mix on vortex and incubate at 55°C for 1.5hours in water bath. Remove the tubes from water bath and allow it to cool to room temperature. Add 580μl of 24 N H_2SO_4 slowly from the sides of the tube, vortex it, incubate at 55°C for 1.5hour in water bath with frequent shaking. Remove the tubes from water bath and allow it to come to room temperature. Then add 3ml of n-hexane to each tube. Cap them and vortex it for 3min vigorously. Centrifuge at 2000rpm for 8min (for layer separation).Collect upper hexane layer in clean glass tube containing 2-3gm of anhydrous Na_2SO_4. Mix and cover it with foil, leave it overnight at -20°C. Dry, hexane under nitrogen evaporation and add 1ml of DCM to the tube, vortex it and wait for a while for Na_2SO_4 to settle. Transfer DCM layer into another clean glass tube, dry under nitrogen. Reconstitute the tube with 1ml DCM. Filter through PVDF 0.22syringe filter in to 2ml vials to run in gas chromatography.

GC operating conditions: Fatty acid composition of the FAME was determined by capillary gas chromatography on SP^{TM} -2560 capillary column-100x0.25mmx0.2μm installed on a gas chromatograph equipped with flame ionization detector and split injector. Initial oven temperature was 140°C, held for 5min,subsequently increased to 240°C at rate of 4°C/min,and then held for 20min.Helium/nitrogen is used as the carrier gas at a flow rate of 0.5ml/min and column head pressure is 280kpa.Both the injector and detector are to be set at 260°C.The split ratio is 30:1.Fatty acids are to be identified by comparing their retention times with reference FAME standards.

GC Analysis

Injection Sequence

a. Inject solvent blank

b. Inject calibration standard(s)

c. Inject sample extract

d. Re-inject the calibration standard at the appropriate level at least after every 20 injections and at the end of the run to verify instrument response.

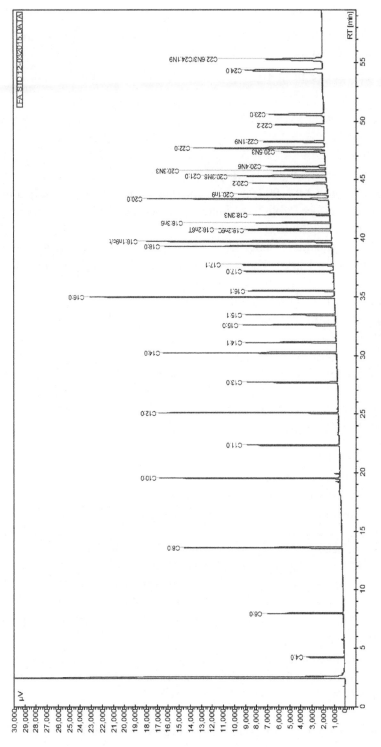

Chromatogram of fatty acid methyl esters

Calculations:

1. Identify the peaks by comparison of their retention time with those of standard fatty acid methyl esters.

2. Calculate the relative proportion (%) of each fatty acid either manually or automatically by software.

References

A.O.A.C. (1975)

Abramson, D. and Blecher M (1964) J.Lipid Research, 5; 628-631.

Church, P.N. and Wood, J.M. (1992) In the manual of manufacturing meat quality, Elsevier Applied Science Publishers Limited, Crown house, Linton road, Barking, Essex, IG11 8JU, England.

Folch, J. Lees, M and Sloane - Stanley G. H. (1957) J. Biol.Chem., 226: 497-509.

Hanahan, D.J. (1960) In Lipid Chemistry" Ed. John Wiley and Sons, Inc. New York.

Hanel, H.K. and Dam, H., (1955) Determination of small amounts of total cholesterol by Tschugaeff reaction with a note on determination of lanosterol. ACTA. Chem. Scand., 9, 677.

J. Lab. Clin.Med., (1953), 41,486 {cited in A Manual of Laboratory Techniques (1983) edited by Raghuramulu, N; Madhavan Nair ,K and Kalyanasundaram, S. Published by National Institute of Nutrition (ICAR) Hyderabad, India.}

Kowale, B.N., Kulkarni, V.V. and Kesava Rao, V. (2008) Methods in meat science. *Pub.* by Jaypee brothers medical publishers (P) Ltd., Ansari Road, Daryagunj, New Delhi.

Marinetti, G.V. (1962) J. Lipid Res. 3: 1-20.

Misra, U.K. (1966) Biochem. Biol. Speri.,8: 125-126.

Misra, U.K. (1969), Biochem.Biol.Speri.8, 128-126.

O'Fallon, J.V., Busboom, J.R., Nelson, M.L., and Gaskins, C. T. 2007. A direct method for fatty acid methyl ester (FAME) synthesis: Application to wet meat tissues, oils and feedstuffs. Journal of Animal Sciences, 85(6):1511-21.

Pie, J.E., Spabis, K. and Seilan, C. (1991) J.Agric. Food Chem.,39: 250-254.

Pie, J.E., Spabis, K. and Seilan, C. (1991) J.Agric. Food Chem.,39: 250-254.

Rouser,G; Bauman, A.J. and Kritcheveski,G. (1961) Am. J. Clin. Nutr., 9: 112-123.

Stahl, E (1965) In Thin Layer Chromatography Laboratory Hand Book, Academic press, Inc, New York, pp 1-27.

Van Handel, E. and Zilversmit D.B., (1957) J. Lab. Clin. Med., 50; 152-157.

Wagner, H. (1967) in Methods for separation and determination of lipids and lipodoses Ed. Schettler, G. pp 190, Springer Verlag, New York

Willemart, R. and Ferdet, J. (1959) Ann. Pharm. France. 17: 366-367.

Willemart, R. and Ferdet, J. (1959) Ann. Pharm. France. 17: 366-367

CHAPTER-5

■ ■ ■

MINERAL ANALYSIS IN MEAT

Minerals are one of most important nutrient for growth and maintenance of the human body. Meat is abundant in several micro and macro minerals required for the body like iron, potassium, phosphrous, sodium, magnesium, zinc and selenium. Further, offals are rich source of iron, zinc and copper. Estimation of the minerals will help in determining the nutritive quality of meat. In research works, mineral estimation is used for evaluating the effect of different feeding regime on mineral content of the meat.

5.1 ESTIMATION OF ASH CONTENT

Weigh 5 to 10 g of the meat sample accurately into a silica crucible. Place the crucible on a clay pipe triangle and heat first over a low flame till all the material is completely charred, followed by heating in a muffle furnace for about 3 to 5 hr at 600°C. Cool it in a desiccator and weigh. To ensure completion of ashing, heat again in the muffle furnace for ½ hr, cool and weigh. Repeat this till two consecutive weights are the same and the ash is almost white or grayish white in colour.

$$\text{Ash content (g/100 g sample)} = \frac{\text{Weight of the ash}}{\text{Weight of the sample taken}} \times 100$$

5.2 MINERAL ESTIMATION BY ATOMIC ABSORPTION SPECTROSCOPIC METHOD

5.2.1 Processing of Sample

Ash Solution

Moisten the ash with a small amount of glass distilled water (0.5 1.0 ml) and add 5 ml of hydrochloric acid to it. Evaporate the mixture to dryness on a boiling water bath. Add another 5 ml of hydrochloric acid and evaporate the solution to dryness as before. Add 4 ml hydrochloric acid and a few ml of water and warm the solution over a boiling water bath and filter into a 100 ml volumetric flask using Whatman No 40 filter paper. Cool and make the volume up to 100 ml and suitable aliquots are used for the estimation of phosphorus, iron and calcium.

Wet Digestion

Organic matter in food materials or solids can be destroyed and oxidised by boiling with sulphuric, perchloric and nitric acids.

1. Weigh about 5 g of minced meat sample in a 100ml Kjeldahl flask.
2. Add 25 ml 3:2:1 (concentrated nitric acid- 60 percent perchloric acid -conc. Sulphuric acid) mixture and shake well. Add few clean glass beads (acid washed)
3. Heat for about 30 min cautiously until the initial vigorous reaction has subsided (dense yellow fumes will be evolved)
4. Heat more strongly for 4 hr until most of the nitrous fumes are removed.
5. Continue heating until white fumes of perchloric acid are evolved. If charring occurs or flask contents tend to become dry, remove from heat, cool and add 5 ml of nitric acid and continue heating for 1 hour.
6. After cooling, transfer quantitatively with deionised water the contents of the digestion flask to 15 ml graduated test tube and make up to the 10 ml mark.
7. The tubes are centrifuged for 30 min at 5000 rpm and transfer the ash solution to acid-washed polythene bottles and store in a cool place prior to analysis by atomic absorption spectrophotometer.

5.2.2 Analysis of Samples using AAS

Representative sample in a suitable liquid from is sprayed onto the flame of an atomic absorption spectrophotometer and the absorption or emission of the mineral to be analysed is measured at a specific wavelength.

Reagents

Stock Solution, 1000 mg / ml :- Dissolve 1 g mineral equivalent of mineral salt in about 10 ml of pure HCl Dilute 1 liter with deionised water.

Working Standard Solutions

Dilute aliquots of the stock solutions with deionised water to make at least 4 standard solutions of each element within the range of determinations.

Procedure

1. The mineral to be determined, standards and blank solutions are aspirated into the flame directly or after suitable dilutions.
2. The optimum operating conditions recommended by the instrument manufacturer should be used.
3. Read at least 3 to 4 ranges of standard solutions before and after sample readings. Flush burner with deionised water between samples and check for zero setting.
4. Prepare calibration curve from the readings of standards.
5. Determine the concentration of samples from the standard graph.

Calculation

$$\text{Ppm mineral} = (\mu g \text{ mineral} / ml) \times \frac{\text{Dilution factor}}{\text{ml aliquot} \times \text{sample weight}}$$

Operation wavelengths (nm)

Cu −324.07, Fe - 248.2, Mg - 285.2,

Mn - 279.5, Zn - 213.9, Cr - 357.9 Ca- 422.7

5.3 ESTIMATION OF CALCIUM CONTENT BY OXALATE METHOD

5.3.1 Estimation of Calcium Content by Oxalate Method

Principle

Calcium is precipitated as oxalate and is titrated with potassium permanganate

Reagents

1. 4 percent ammonium oxalate solution
2. Dilute ammonia solution (2 ml of liquor ammonia + 98 ml water)
3. 1 N H_2SO_4
4. 0.01 N potassium permanganate solution

5.   0.01 N Oxalic acid: Sodium oxalate is dried in an oven at 100 to ±05°C for 12 hr. Exactly 0.67 g is dissolved in redistilled water, 5 ml concentrated H_2SO_4 is added and solution made up to one liter after it has cooled down.

Standardisation of potassium permanganate solution: 25 ml of 0.01 N oxalic acid is transferred to an Erlenmeyer flask. Add 1 ml of concentrated H_2SO_4. Warm to about 70°C and titrate against $KMnO_4$ solution, till the faint pink colour remains.

Procedure

Take 2 ml sample into a 15 ml centrifuge tube. Add 2 ml of distilled water and 1 ml of 4 percent ammonium oxalate solution and mix thoroughly and leave overnight. Again mix the contents and centrifuge for 5 min at 1500 rpm, discard the supernatant and drain the tube. Wash the precipitate with 3 ml of dilute ammonia. Centrifuge and drain supernatant. Repeat once more to ensure complete removal of ammonium oxalate. Dissolve the precipitate in 2 ml of 1 N H_2SO_4. Place the tube in boiling water-bath for 1 minute and titrate against 0.01 N $KmnO_4$ solution to a definite pink colour persisting for at least 1 minute.

Calculation

1 ml of 0.01 N $KMnO_4$ is equivalent to 0.2004 mg of calcium.

mg of calcium per 100 ml of ash solution $=(X-b) \times 0.2004 \times 100/2$

X = ml of 0.01 N $KMnO_4$ required to titrate sample

b = ml of 0.01 N $KMnO_4$ required to titrate blank (2 ml of H_2SO_4)

If the normality of $KMnO_4$ is N , the value obtained in the above formula should be multiplied by the factor N/ 0.01

5.3.2 Estimation of Calcium by Colourimetric Method

Principle

Calcium forms a colour complex with the O-cresopthalein dye, which is made more specific in the presence of 8-quinolinol.

Reagents

1.   Standard $CaCl_2$ – 100 mg Ca^{++} / ml prepared by dissolving required amount of anhydrous $CaCO_3$ in dilute HCl and making up to a definite volume with water.

2. Ammonium chloride buffer (pH 10.5) – dissolve 0.24 g NH_4Cl in 100 ml of 5 percent NH_4OH solution (v /v).

3. O- cresophthalein solution , 0.1percent - 100 mg of O-cresopthalein dissolved in 28 ml of ammonium chloride buffer and diluted to 100 ml with water.

4. 8-quinolinol – 1 percent solution in absolute ethanol.

Procedure

Take different aliquots of $CaCl_2$ solution with Ca^{++} content varying from 2 to10 mg in graduated stoppered tubes and make volume to 5.5 ml with water. Add 1 ml of 8-quinolinol and 2.5 ml of ammonium chloride buffer and adjust volume to 9.0 ml. Add O-cresophthalein and mix. Take water as blank and read at 565 nm.

Calculation

$$\frac{\text{OD unknown} \times \text{concentration of standard}}{\text{OD standard} \times \text{aliquot volume in ml}} \times 100$$

(Anal.Biochem 1967)

5.4 ESTIMATION OF PHOSPHORUS CONTENT

Principle

In the determination of acid-soluble, lipid and total phosphorus, organic matter is destroyed by digestion with H_2SO_4 (or fuming nitric acid). The phosphate containing solutions are treated with molybdic acid to produce phosphomolybdic acid. This is reduced by the addition of 1,2,4 amino naphtholsulphonic acid reagent giving a blue colour, the intensity of which is proportional to the amount of phosphate present.

Reagents

1. 10 percent TCA solution in water
2. 10N and 5N H_2SO_4
3. 30 percent H_2SO_4
4. Aminonaphtholsulphonic acid reagent (ANSA) Take 195 ml of 15 percent sodium bisulphate solution in a glass stoppered cylinder and add 0.5g of 1,2,4 aminonaphtholsulphonic acid to it followed by 5 ml of 20 percent sodium sulphite. It is stoppered and shaken until the powder is dissolved. If the solution is not complete, more of sodium sulphite, 1 ml at a time is added

with shaking. The solution is then transferred to a brown glass bottle and stored in the cold. This solution is usable for four weeks.

5. Standard Phosphate Exactly 351 mg of pure dry monopotassium phosphate (KH_2PO_4) is dissolved in water and transferred quantitatively to a one liter volumetric flask. Ten ml of 10 NH_2SO_4 is added and the solution diluted to the mark with water and mixed. This solution contains 0.4 mg of phosphorus in 5 ml. It is stable indefinitely.

6. 2.5 percent Ammonium molybdate: 2.5 g of reagent grade ammoniummolybdate is dissolved in water, transferred to 100 ml volumetric flask, filled to the mark and mixed.

Procedure

For photometric measurement suitable aliquot from ash solution, series of standard solution and a blank are digested with 2.5 ml of 5 M H_2SO_4 on a microburner. After evaporation is complete and the mixture turns brown or black with no further change, remove the tubes, cool slightly and add 1 drop of 30 percent H_2O_2 and 1 or 2 drops of fuming nitric acid. Continue heating until the contents of the tube become colourless. Transfer the contents to 25-ml volumetric flask quantitatively with the help of distilled water. Add 2.5 ml of 2.5 percent Ammonium molybdate solution followed by 1 ml of aminonaphtholsulphonic acid reagent to each flask, mix well and allow to stand for 5 min. Read the intensity of the colour at 660 nm setting the photometer to zero with blank. Plot the standard curve and read the unknown value from it (Fiske and Subba Row, 1925).

5.5 ESTIMATION OF IRON CONTENT

Introduction

Iron content of meat depends primarily on the myoglobin content, hence, meat with higher myoglobin content will have higher iron content. Reported values of iron content (mg /100g) in raw beef, mutton chop, pork, and bacon are 2.3, 1.0, 1.4 and 0.9 respectively. Free iron released on cooking of meat is responsible for the catalyzing oxidative rancidity of fats. Addition of EDTA (disodium) to meat has been successfully tried to bind with the iron and delay the oxidative rancidity in meats.

Equipment

Fine analytical balance, high-speed blender (15000 rpm), Spectrophotometer.

Reagents

Concentrated Nitric acid, 70 percent Perchloric acid, Saturated Potassium per sulphate Sodium tungstate (10 percent) Potassium thiocyanate (3N).

Procedure

Digest 5 g meat with 10 ml of concentrated Nitric acid by heating. Add 5 ml of 7 percent perchloric acid. Boil with caution, gently to drive out NO_2 fumes until the solution is clear and colourless. Cool, make volume to 100 ml with distilled water. To 5 ml aliquot add 2 ml saturated Potassium persulphate mix and dilute to 25 ml with distilled water. Add 2 ml of sodium tungstate (10 percent). Wait for 5 minutes. Make volume to 50 ml mix well. Filter through Whatman No.1. Pipette 5 ml add 0.25 ml saturated potassium persulphate and 1 ml of potassium thiocyanate (3N). Read at 480 nm within 30 min. in a spectrophotometer.

For standard take 2.5 ml of Fe Standard (0.1 mg/ml) to this add 2 ml of conc. H_2SO_4 and 2 ml Potassium persulphate. Dilute to 25 ml and add 2 ml (10percent) Sodium tungstate, make volume to 50 ml. Mix well and filter. To 5 ml of aliquot add 0.25 ml saturated potassium per sulphate and 1 ml of 3 N Potassium thoicyanate. Read at 480 nm. For blank take 5 ml distilled water, add 0.25 ml saturated potassium per sulphate and 1 ml 3 N Potassium thiocyanate.

Calculation

$$\text{Mg of Fe / g of meat} = \frac{\text{O.D. of unknown} \times \text{Conc. of standard}}{\text{O.D. of standard}} \times 4$$

(Wong, 1928)

5.5.1 Qualitative Estimation of Iron Content in Meat

Iron may cause discolouration in meat products. A common source of iron contamination is the water supply. A simple and quick test will suffice and can be performed for trouble- shooting purposes anywhere. Hydrochloric acid when spread over the sample will react with potassium ferrocyanide and if iron is present, blue colour will appear in a few minutes.

Reagents

1. 4N Hydrochloric acid
2. 5 percent solution of potassium ferrocyanide

Procedure

1. Take a slice of suspect meat product, add 1to 2 drops of the hydrochloric acid and spread it out with glass stirring rod.

2. After about one minute add 1to2 drops of the potassium ferrocyanide solution onto the same location. Within 2 to3 minutes the blue colour will appear in the presence of iron. The darker is the blue colouration, higher the ppm Fe content.

   Discard all meats tested and store these reagents safely from the processing area. Preferably in the laboratory same procedure may be used for emulsions, brine, pickle solutions and other matter suspected to be contaminated by iron.

5.5.2 Quantitative estimation of Iron Content in meat by Colourimetric Method

Principle

Ferrous iron reacts over a wide range of pH with 1,10-phenolpthaline to form a deep red colour complex. Total iron is obtained in the sample by using acid to dissolve precipitated iron and by applying hydroxylamine HCl to reduce ferric iron to the ferrous state. When the test colour is developed at the buffered pH of 4.5 to 5.0, the colour is very stable. This method detects as low as 0.01 ppm of iron (Fe) and as high as 2.0 ppm, depending how the solutions are diluted.

This test is free from much interference. Polyphosphates, which would normally interfere, are reverted to ortho-phosphate by the initial boiling step in acid. Boiling also removes cyanides and nitrites that would interfere. Zinc, chromium, copper, cobalt if concentration is less than 5 ppm will also not interface in test.

Reagents

1. Hydrochloric Acid Concentrated reagent grade hydrochloric acid (HCl-specific gravity 1.19).

2. Hydroxylamine hydrochloride solution: Dissolve 10 g of reagent grade hydroxylamine hydrochloride in 100 ml of distilled water.

3. Acetate Buffer solution: Dissolve 250 g of reagent grade ammonium acetate ($NH_4C_2H_3O_2$) in about 700 ml of distilled water. Add exactly 70 ml reagent grade glacial acetic acid (specific gravity 1.05) to the solution and dilute to one liter with distilled water.

4. Phenolphthalein indicator -Dissolve 0.5 g of 1,10-phenolpthaline monohydrate in a mixture of 100-ml ethyl alcohol and 100 ml distilled water.

5. Ferrous Ammonium sulphate, reagent grade, fine crystal, (MW 392.14).

Standard Iron Solution

First prepare a stock solution by dissolving 0.7022 g ferrous ammonium sulphate ($FeSO_4$ $(NH_4)2SO_46H_2O$) in about 700 ml distilled water containing 10 ml of concentrated reagent grade sulfuric acid. Dilute to one liter in a volumetric flask. One ml of this stock solution contains 0.1 mg Fe. Prepare Standard Iron Solution as needed by diluting 100 ml of the stock solution to one liter with distilled water. One ml of standard iron solution contains 0.01 mg Fe or is equivalent to 0.2 ppm Fe when diluted to 50 ml and 0.1 ppm when diluted to 100 ml.

Procedure

1. Weigh 1 g of meat sample and ash the sample in porcelain dish.

2. Cool and add a little water, 1 ml of concentrated hydrochloric acid and 1 ml of hydroxylamine HCl. Dissolve the ash and transfer to a flask, washing the porcelain dish with hot water making altogether about 50 ml solution.

3. Add several glass beads to flask to prevent bumping and boil solution until volume is reduced to 15-25 ml.

4. Cool to room temperature and add 1 ml of phenolphthalein solution and mix.

5. Add 10 ml of acetate buffer solution to the flask, transfer to 50 ml, graduated cylinder and dilute to mark with distilled water. Mix well. The final test will show result on basis of 50 ml dilution.

6. Allow solution to stand for 15 minutes to develop full colour, set up spectrophotometer on zero with blank solution (water, phenolphthalein and acetate buffer) and measure optical density at 510 nm.

    Calculate or read from standard curve and report as Fe in ppm. If the standards are available in Nessler tubes, compare the sample and report on the basis of the colour.

References

Anal.Biochem , 20, 155 (1967).

Church, P.N. and Wood, J.M. (1992) In the manual of manufacturing meat quality, Elsevier Applied Science Publishers Limited, Crown house, Linton road, Barking, Essex, IG11 8JU, England.

Fiske, C.H. and Subba Row (1925) J.Biol. Chem, 66 : 375-400.

Koniecko, E.K. (1979); In Hand book for Meat Chemists, pp 68-69.Avery Publishing Group Inc. Wayne, New Jersey, USA.

Kowale, B.N., Kulkarni, V.V. and Kesava Rao, V. (2008) Methods in meat science. *Pub.* by Jaypee brothers medical publishers (P) Ltd., Ansari Road, Daryagunj, New Delhi.

Raghuramulu, N; Madhavan Nair and Kalyanasundaram,S.(1983) In A Manual of Laboratory Techniques. National Institute of Nutrition, Indian Council of Medical research, Jamal-Osmania, Hyderabad-500007.

Wong S.Y. (1928) J. Biol. Chem 77 : 409-412.

CHAPTER - 6

■ ■ ■

TESTS FOR DETERMINING FUNCTIONAL PROPERTIES OF MEAT

6.1 MEASUREMENT OF pH

pH influences the quality of meat i.e. colour, tenderness, flavour, water binding properties and shelf life. Measurement of pH can, therefore, reveal the quality of meat in a carcass and decides whether the meat is suitable for the manufacture of good quality products.

The pH of muscles falls after slaughter due to production of lactic acid from glycogen via Embden-Meyerhof pathway to reach an ultimate level of 5.5 to 5.7 followed by steady rise, which accelerates as decay, begins. This ultimate pH varies indifferent carcasses and muscles of the same species. Struggle of animals just before slaughter results in markedly lower initial pH and early passing off of rigor mortis.

Principle

The pH represents the acidity or alkalinity of an aqueous solution. The pH scale expresses the hydrogen or hydroxyl ion concentration in sample from 1.0 (highly acidic) to neutral (7.0) and to 14.0 (highly alkaline.)

Apparatus and requirements

1. pH meter: Regular or digital, equipped with appropriate electrodes.
2. Standard buffer solutions, of pH 4, pH 7 and pH 9.1, for various tests.
3. Waring Blender
4. Filter paper: Whatman
5. Measuring pipettes
6. Analytical scale

Procedure

1. Weigh 10 g of the meat or meat product sample.
2. Add 90 ml. of distilled or deionized water and blend for a minute.
3. Standardize the pH meter by using standard buffer solution.
4. Observe the thermo - compensator for buffer and sample.
5. Set the instrument to a buffer value, which is closer to sample to be analysed. Wipe the electrodes and place in the beaker with distilled or deionized water.
6. Insert electrodes again into sample and wait until indicator or digital control slightly fluctuates or is stable. About 5 minutes of time is sufficient.
7. Observe the instructions included to operate pH meter.
8. When test is completed, wipe the electrodes and place them into a beaker with distilled water (AOAC, 1975).

6.2 MEASUREMENT OF WATER HOLDING CAPACITY

Introduction

Many physical properties of meat such as the colour, texture and firmness of raw meat and juiciness and tenderness of cooked meat are partially dependent on water holding capacity (WHC). Muscle proteins are responsible for the binding of water in meat. In beef homogenate with 60 percent added water, the total water holding capacity is distributed on different components viz. structural proteins 65 percent, water-soluble proteins 5 percent and water non-soluble proteins 30 percent. The WHC is minimum at isoelectric point of muscle proteins and increases on either side of pH due to unfolding of peptide chains there by loosening of meat structure and increasing the net charge on meat proteins. However, pH and WHC of beef are not correlated within the pH range of 5.4 to 5.8 to the total moisture and protein content of meat even though total moisture is highly correlated to protein content. Meat with high intramuscular fat has higher WHC due to loosened microstructure of tissue and more and more water is held in the immobilised state. Influence of connective tissue on WHC is not clear. Besides species, sex, age, grade, breed and muscle also influence the WHC.

6.2.1 Estimation of WHC by Filter Paper Press Method

Principle

Water holding capacity (WHC) is the ability of meat proteins to retain its own water content when subjected to pressure treatment.

Equipment

Fine analytical balance, filter paper

Procedure

Weigh two Whatman no. 1 filter papers. Weigh around 500 mg of minced meat sample. Place meat sample in between the centre of two weighed filter papers. Place this on a rigid, flat surface by keeping polyethylene sheet above and below. Apply pressure (40 psi) i.e. 2.81 kg on it for 5 min. Weigh the meat flake. Dry the filter papers and weigh.

Calculation

WHC is expressed as percent. Determine actual weight of meat flake by adding the weight of meat flake after pressing and subtraction of weights of filter papers before and after (this gives the amount of proteins attached to filter paper)

$$\text{WHC as percent} = \frac{\text{Weight of meat flake}}{\text{Sample weight}} \times 100$$

(Whiting and Jenkins, 1981)

6.2.2 Estimation of WHC by Centrifugation Method

6.2.2.1 Estimation of WHC using Distilled Water

Principle

Water holding capacity (WHC) is the ability of meat proteins to retain its own water content during various stages of processing.

Equipment

Fine analytical balance, refrigerated centrifuge

Procedure

Weigh about 5 g of minced meat sample. Place meat sample in centrifuge tube; add 10 ml of cold distilled water and stir for 5 min with glass rod. Keep at 0°C for 15 min and centrifuge at 10000 rpm for 10 min. discard the supernatant. The quantity of water retained by the sample is calculated by subtracting initial weight from final. The WHC is expressed as percent.

Calculation

$$\text{WHC percent} = \frac{(\text{Final weight - Initial weight})}{5} \times 100$$

(Wardlow *et al.*, 1973)

6.2.2.2 Estimation of WHC using NaCl Solution Method

Weigh 15 g minced meat, add 22.5 ml of 0.6 M NaCl, stir for one minute and allow it to stand for 15 minute at 4°C. Stir and centrifuge at 10000 rpm for 15 minutes in a refrigerated centrifuge at 4°C. Measure the volume of supernatant. WHC is expressed as percent.

Calculation

$$\text{WHC percent} = \frac{(\text{Final volume} - \text{Initial volume})}{15} \times 100$$

(Hamm, 1960)

6.3 MEASUREMENT OF EMULSIFYING CAPACITY

Introduction

Emulsifying capacity (EC) is the ability of proteins to emulsify or hold a maximum amount of fat under specific conditions. EC and pH are highly correlated. Adjustment of pH of post rigor poultry muscle to that observed in prerigor state restored the original high EC of prerigor muscle. The EC of meat proteins in the descending order is actin, myosin, actomyosin, sarcoplasmic proteins has been reported. The EC is minimum at isoelectric point of muscle proteins and increases on either side of pH. The EC is also influenced by the concentration of salt soluble proteins, temperature, speed of mixing and bacterial load.

Equipment

Fine analytical balance, high-speed blender (15000 rpm)

Procedure

Weigh around 25 g of minced meat sample add 100 ml of 1 M NaCl, keep overnight at 0 to 4°C. Blend at 13000 rpm to prepare the slurry. Weigh 2.5 g slurry and add 37.5 ml 1 M NaCl, mix well. Initially add 50 ml refined edible oil and blend in a plastic emulsifying jar at 13000 rpm, simultaneously running down measured quantity of oil @ 0.8 ml / sec. Perfect emulsion gives a look of a homogenous curd. The sound of the blender changes at the point of emulsion breakdown and gives the appearance of broken curd. Note the amount of oil required at break point of emulsion.

Calculation

The EC is expressed as the amount of oil emulsified by one gram of meat protein.

Emulsifying Capacity (protein estimation)

Equipment and reagent

Fine analytical balance, vortex mixer, Folin Ciocalteau reagent and 1 M NaCl.

Procedure

Weigh 2 g of minced meat sample add 100 ml of 1 M NaCl, homogenise in vortex mixer for 20 min, keep at 0 to 4°C for 30 min for extraction of salt soluble proteins. Centrifuge for 5 min at 6000 rpm. Filter through Whatman no.1. Pipette 0.05 ml filtrate in a test tube to be used for protein estimation. Take 30 ml filtrate; add 10 ml of refined oil. Make emulsion by keeping the temperature below 10°C. Settle it in a refrigerator for 15 min. Centrifuge for 1 min at low speed. Take 0.1 ml of aliquot from liquid portion for protein estimation by Lowry's method.

Calculation

EC is expressed as the amount of oil emulsified by one gram of meat protein (Swift *et al.*, 1961).

6.4 MEASUREMENT OF MUSCLE FIBRE DIAMETER

6.4.1 Estimation of Muscle Fibre Diameter by Direct Homogenization Method

Introduction

The essential structural unit of all muscles is the fibre. Fibres are long, narrow multinucleated cells, which may stretch from one end of the muscle to the other end and may attain a length of 34 cm, although they are only 10-100 mm in diameter. In healthy animals the muscle fibres diameter differs from muscle to muscle, between species, breeds and sexes. Size / diameter are increased by age, plane of nutrition, training, by degree of postnatal development in the body weight rather than by body weight itself and by oestradiol administration. The rate of increase in size of the muscle fibre declines as an animal approaches maturity. Muscle fibres of male are usually larger than those of female or castrates while those of male castrates being intermediate in fibre size.

As the animal grows older the total number of muscle fibres will decrease and remaining fibres will become larger. Thus, skeletal muscles in old age contain fewer fibres, which are larger than those at maturity.

Procedure

Weigh about five g of muscle from cross-sectional surface sample and cut into small cubes and homogenize in a blender twice for 15 sec periods at low speed interspaced with a five seconds resting interval in a solution containing 0.25 M sucrose and 1 mM EDTA to produce a slurry. Transfer one or more drops of the slurry on to a microscope slide and cover with a cover slip. Examine the suspension directly under a light microscope using a calibrated micrometer and the muscle fibre diameter is measured as the mean cross-sectional distance in micrometer between the external borders of the sarcomere of 20 randomly selected muscle fibres (Jeremiah and Martin, 1977).

6.4.2 Estimation of Muscle Fibre Diameter by Formalin Fixing Method

Muscle fibre diameter was determined as described by Tuma, Venable, Wuthier, & Henrickson (1962). A small core (1.00 cm) of muscle tissue was fixed in formal saline for 24 h and was blended in a micro blender at low speed for 30 s. A drop of the homogenate was placed over a glass slide, covered with cover slip and observed under a microscope with 10 X eyepiece containing a calibrated micrometer. The diameter of 21 fibres was measured and the average muscle fibre diameter was expressed in microns.

Composition of 10% formal saline

40% formalin	-	10 ml
Dist. Water	-	90 ml
Sodium acetate	-	2 gm
Sodium chloride	-	2 gm

Calculation of Muscle Fibre Diameter

Under 40X

12 digits of ocular micrometer (OM) = 5 digits of stage micrometer (SM)

Each digit of SM = 0.01 mm or 10 μm

Each digit of OM = 50/12 which is 4.16 μm

Each digit of OM = 4.16 μm

Under oil immersion

5 digits of SM = 37 digits of OM

i.e 37 digits = 50 μm

1 digit of OM = 50/37 which is 1.35 μm

Each digit of OM = 1.35 μm

(Tuma, 1962)

6.5 Measurement of Sarcomere Length

Introduction

The unit of the myofibril between two adjacent Z lines is called sarcomere. The sarcomere is the repeating structural unit of the myofibril and it is also the basic unit in which the events of the muscle contraction relaxation cycle occur. Sarcomere length is not constant and its dimensions as well as those of I band are dependent on the state of contraction at the time muscle is examined. In mammalian muscle at rest, a sarcomere length of 2.5 mm is rather typical. Sarcomere length is affected by conditioning, cooking, post-mortem temperature and mode of suspension of carcasses. Sarcomere length can be used for prediction of both longitudinal growth potential of a muscle and as indicator of meat tenderness during its conversion to meat.

Procedure

Weigh about five g of muscle from cross-sectional surface sample and cut into small cubes and homogenise in a blender twice for 15 sec periods at low speed in a blender in 30 ml of 0.25 M Sucrose solution for 60 sec. Transfer a drop of the homogenate on to a glass slide and examine under a light microscope under oil immersion objective. If the fibres are not sufficiently broken apart, homogenise the sample for an additional period of 20 to 30 seconds. The sarcomere length is measured using a calibrated ocular micrometer. Measure the sarcomere length of twenty myofibrils randomly, calculate the average and express in micrometers (Cross, 1979).

6.6 MEASUREMENT OF MYOFIBRILLAR FRAGMENTATION INDEX

Introduction

Myofibrillar Fragmentation Index (MFI) is related to the degeneration of myofibrillar proteins during post mortem storage of meat. A potential practical application of the MFI would be to predict the level of tenderness in fresh and cooked steaks. The MFI values coincide with W-B shear force values, sensory values and microscopic examination of muscle fibres. MFI increases during refrigerated storage of muscle. It has been reported that MFI accounted for more than 50 percent of the variation in loin steak tenderness and that MFI is more important effectors of tenderness in loin steak than collagen solubility or sarcomere length.

Reagents

1. Isolating medium: - 100 mM potassium chloride (7.456g), 20 mM potassium phosphate (dibasic, 2.484 g), 1mM EDTA (372.24 mg), 1mM $MgCl_2$ (203.31 mg), 1mM sodium azide (656 mg) dissolved in a liter of distilled water.
2. Biuret reagent: - Weigh 6 g sodium potassium tartrate and 1.5 g CuSO4, to this add with constant stirring in 300 ml of 10percent NaOH. Dilute to 1000 ml with distilled water.

Procedure

Homogenise 4 g of minced meat in 40 ml of chilled isolating solution for 30 sec. Centrifuge at 5000 rpm for 15 minutes and decant the supernatant. Resuspend the sediment in 40 ml of isolating medium by using a stirring rod. Centrifuge again as above and decant the supernatant. Resuspend the sediment in 10 ml of isolating medium, and filter through polyethylene strainer (18 mesh) to remove connective tissue and debris. Wash the strainer with another 10 ml of isolating medium to facilitate passage of myofibrils through the strainer. Determine the protein concentration in the filtrate by Biuret method.

Adjust the protein concentration of the extract to 0.5 mg / ml by diluting with isolating medium. Again determine the protein concentration by Biuret method. Dilute the myofibril suspension and pour it into cuvette. Measure optical density at 540 nm.

Calculation

MFI of sample = OD × 200 (Oslon *et al.*, 1976 and Gornall *et al.*, 1949)

Myofibrillar Fragmentation Index

Freeze breast muscle cubes of 7mm x 7mm in liquid nitrogen. Weigh 10 g of such cubes and transfer to a 250 ml vertis stainless steel homogenisation cup containing 50 ml of cold solution of 0.25 M sucrose and 0.02 M potassium chloride. Thaw the muscle cubes for 5 min. Homogenise the sample in a vertis homogeniser 45 (Virtis company, Gardiner, New York, USA) for 40 sec at 45,000 rpm with two virtis macro stainless steel blades aligned and positioned 1 mm below the surface of the solution and set in a reverse position. Filter the homogenate through a pre-weighed 250 m stainless steel wire cloth screen in the filtration unit stirring with a glass rod to expedite the filtration process. Blot the residue on a Whatman no.3 filter paper. Weigh the residue taken at the end of 40 min of drying in a hot air oven chamber maintained at 40°C. The fragmentation index is obtained by multiplying the weight of the fraction remaining on the screen in g by 100 (Davis *et al.*, 1980).

6.7 MEASUREMENT OF COLOUR OF MEAT

6.7.1 Measurement of Colour using Lovibond Tintometer

Lovibond Tintometer is used for the examination and measurement of a full range of coloured sample- either liquid or solid by transmitted or reflected light. It comprises of observation cabinet, graduated Lovibond glass colour filters, and two standard white references. The sample holders are attached to the rear of the equipment. Meat sample is passed through 4-mm sieve to prepare it for putting in to the sample holders. Having placed the sample in position such that it can be seen in the left hand field of the viewing tube, slide the triangular knobs (which control the coloured filters) towards the right adjusting the red, yellow and blue in correct proportion until a perfect colour match is obtained. Record the values of the slides as shown in the indicator apparatus. The Lovibond unit of colour is an arbitrary one but through long and worldwide use, it has received international acceptance. It describes the appearance of colour in very simple words. Many people record the result on 3 colour basis - red, yellow and blue. Some others use all the 6 colours red, orange, yellow, green, blue and violet. A meat sample is described as dull brown when all the three of the colour slides, red, yellow and blue together are needed to make a match. The value of blue colour reflect dullness while comparing two meat samples the value of red colour, in exact can be obtained by subtracting the value of yellow colour.

Interpretation		
	Stored meat	Fresh meat
Red	2.9	1.9
Yellow	1.8	0.6
Blue	0.8	0.4

6.7.2 Measurement of Colour by Using Hunter Colour Lab

Colour is the first attribute of meat to be evaluated by the consumers and it is of utmost importance especially for fresh meat at the retail level. Colour of the meat is produced by myoglobin protein present in meat and the concentration and status of myoglobin decides the redness of meat. Generally preferred meat colour by consumers are bright red colour in fresh meat, brown or gray-color in cooked meats and pink colour in cured meat. Poultry meat with its lower myoglobin content (<0.5 mg/g meat) is also known as white meat. Meat colour depends on number of ante mortem factors, such as age, the physiological function of the muscle, the nutritional status of the bird and the dietary regime.

Principle of colour measurement

CIE L×a× b×color based method is the most commonly used method for colour evaluation of meat. L, a, b based colour scores are based on opponent colour theory which is conceptually based on the assumption that human eye perceives colour as pair of opposites. L scale compares lightness vs darkness, a scale compares redness vs greenness and b scale compares yellowness vs blueness of the product.

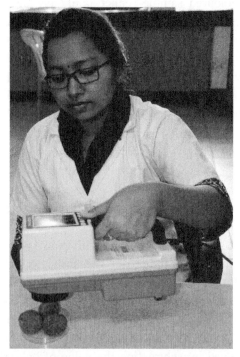

Hunter Colour Lab   **Measurement of colour using Hunter color Lab**

Procedure of colour measurement of meat using Hunter Colour Lab

Meat color analysis is performed using a Hunter lab Miniscan XE Plus colorimeter (Hunter Associates Laboratory Inc., Reston, VA, USA) with 25 mm aperture set for illumination D65, 10° standard observer angle. CIE L^* (lightness), a^* (redness) and b^* (yellowness) are measured on the surface of meat from five randomly chosen spots. *Hue angle* [Tan^{-1} (b*/a*)] and *chroma* (a*^{2} + b*^{2})$^{1/2}$ are calculated according to Hunter and Harold (1987).

6.7.3 Estimation of Total Pigment

Introduction

The pigments in meat consist largely of two proteins, hemoglobin, the pigment of the blood and myoglobin, the pigment of the muscles. In well-bled muscle tissue, myoglobin constitutes 80-90 percent of the total pigment and is much more abundant than hemoglobin. Other pigments such as the catalase and cytochrome enzymes may be present, but their contribution to meat colour is minor.

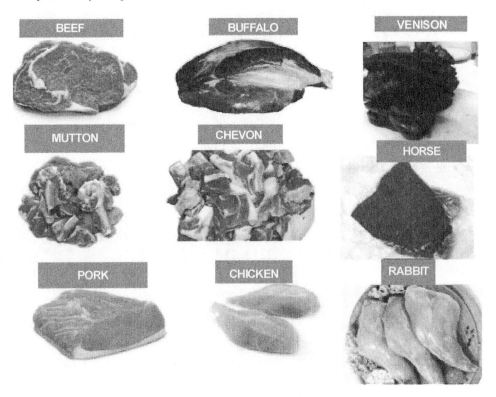

Myoglobin quantity varies with species, age, sex, muscle type and their physiological activity. Beef, chevon, mutton has more myoglobin than pork, veal, fish and chicken. Muscle to muscle differences in myoglobin content are due to the type of muscle fibre i.e. white and red. The chemical state of myoglobin is affected by exposure to atmosphere, oxygen, addition of preservatives such as nitrite, method of cooking, packaging, etc.

Procedure

Take 50 g of meat (free of excess connective tissue and fat) cut into small pieces and homogenise to get a fine homogenate.

Weigh 10 g of this homogenate and add 10 ml of distilled water in a 50 ml beaker and keep overnight in a refrigerator. Centrifuge for 15 min at 5000 rpm in a cold room and filter the supernatant through Whatman No.1 paper. Re-extract the residue with 40 ml of distilled water for 30 minutes, centrifuge and filter as above. Combine the two extracts in 50 ml volumetric flask and make up the volume. Take 20 ml of the filtered clean aliquot and to this add 0.005 g of potassium ferricynide (0.1 ml of 50 mg / ml) and 0.001 g of potassium cyanide (0.1 ml of 10 mg / ml). Centrifuge at 5000 rpm for 5 minutes and read at 540 nm.

Calculation

Total pigment (mg / g wet tissue) = OD × K / sample weight

The proportionality constant K = 17000 × volume in liter / E

E = milimolar extinction coefficient of myoglobin is 11.3.

The volume of the total sample solution equals 0.50 liter. The molecular weight of myoglobin is 17000 mg / mM.

Hence K= 75.2212

Total pigment (mg / g) = OD × 75.2212 / 10 = OD × 7.52

6.7.4 Estimation of Myoglobin Content

A 20 ml aliquot of the extract remaining from the total pigment determination can be used. To this add 2 ml of 0.5 M phosphate buffer (pH 7.1) followed by 5.5 ml of saturated basic lead acetate, to precipitate proteins. The temperature must be closely controlled because at the temperature higher than 38°C, myoglobin also precipitates and at temperature below 6°C the protein precipitation will be incomplete. Filter the supernatant through Whatman filter paper. To the 15 ml filtrate add 6.9 g of mono and dibasic potassium mixture to bring the phosphate concentration to 3 M and pH to 6.6. Addition of phosphate increases the volume to 17 ml. Centrifuge at 5000 rpm for 15 min. at below 2°C. Filter through Whatman paper. To the filtrate add Potassium ferricynide (0.1 ml of 4 mg/ml) to the concentration of 0.6 mM/lit to convert myoglobin to metmyoglobin. Then add Potassium cyanide (0.1 ml of 10 mg /ml) to the concentration of 0.8 mM / lit which will convert metmyoglobin to cynametmyoglobin. Centrifuge at 5000 rpm for 15 min and read the optical density of filtrate at 540 nm.

Calculation

Myoglobin (mg / g wet tissue) = OD × K / sample weight (10g)

The proportionality constant K = $\dfrac{17000 \times \text{volume of aliquot in lit} \times \text{dilution factor}}{E}$

E = milimolar extinction coefficient of myoglobin is 11.3.

K = 17000 × 0.017 × 2.5/ 11.3

Myoglobin (mg / g) = OD × 6.39

(Aragnosa and Henrickson, 1969)

6.7.5 Preparation of myoglobin for *In Vitro* studies

Myoglobin (MW ~ 17,000 Da) can be purified readily from skeletal or cardiac muscle of meat producing animals or poultry. It is a robust protein and its red color permits easy visualization of its progress during chromatography. A variety of purification procedures for myglobins have been published; a straightforward approach adapted from both Faustman and Phillips (2001) and Naveena *et al.* (2009), which provides substantial yields of myoglobin using relatively inexpensive equipment, is provided.

MATERIALS:

• Meat samples trimmed of visible fat and connective tissue

• Blender

• Homogenization buffer; 10 mM Tris-HCl, 1 mM EDTA, pH 8.0, 4°C for mammalian livestock species or 20 mM ammonium bicarbonate buffer; pH 9.0, 4°C for poultry species.

• Cheese cloth

• Centrifuge

- pH meter
- Ammonium hydroxide
- Ammonium sulphate
- Stir plate
- Dialysis tubing (MWCO 12-14 kDa)
- Dialysis buffer (5 mM Tris-HCl, 1 mM EDTA, pH 8.0 or 10 mM ammonium bicarbonate buffer; pH 9.0, 4°C)
- Sephadex G-100 or Sephacryl HR-200 chromatography column (120 x 2.5 cm)
- Chromatography elution buffer (5 mM Tris-HCl, 1 mM EDTA, pH 8.0 or 20 mM ammonium bicarbonate buffer; pH 9.0, 4°C)
- Diethylaminoethyl (DEAE) Cellulose
- Peristaltic pump
- Fraction collector

METHODOLOGY:

1. Homogenize 150 g meat (diced) in 450 mL homogenization buffer.
2. Centrifuge the homogenate at 5,000 x g for 10 minutes. Discard the precipitate and pool the supernatants. Adjust the pH of the resulting supernatant to 8.0 (or 9.0 for poultry meats) using ammonium hydroxide.
3. Filter the homogenate through 2 layers of cheesecloth to remove lipid and connective tissue particles.
4. For Bovine and Porcine-Bring the extract from 0% to 70% ammonium sulphate saturation (436 g/L); For Ovine and Poultry – Bring the extract from 0% to 50% ammonium sulphate saturation (323 g/L). Adjust the pH to 8.0 (or 9.0 for poultry meats) and stir for 1 hour.
5. Centrifuge the solution at 18,000 x g for 20 minutes to remove precipitated proteins. Discard the precipitate and pool the supernatants.
6. For Bovine and Porcine – Bring the supernatant from 70% to 100% ammonium sulphate saturation (209 g/L); For Ovine and Poultry – Bring the supernatant from 50% to 100% ammonium sulphate saturation (323 g/L). Adjust the pH to 8.0 (or 9.0 for poultry meats) and stir for 1 hour.
7. Centrifuge the solution at 20,000 x g for 1 hour. Discard the supernatant and transfer the precipitated myoglobin to dialysis tubing (MWCO 12 – 14 kDa).
8. Dialyze the myoglobin against dialysis buffer (1 vol protein: 10 vol buffer) 3 times, changing the dialysis buffer every 8 hours.

9. Filter solution through 0.45 μm filter using syringe. If desired, concentrate myoglobin solution using Centricon concentrators.

10. Apply the dialysate to Sephadex G-100 or Sephacryl HR-200 gel filtration column which has been equilibrated with chromatography elution buffer. Resolve the myoglobin extract with chromatography elution buffer at a flow rate of 60 mL/hr. Hemoglobin will elute first as a pale red/brown band. Myoglobin will follow as a readily visualized dark red band.

11. After gel filtration chromatography if the myoglobin is not in its pure form (single band, as evidenced by SDS-PAGE) it can be further purified using DEAE Cellulose anion exchange chromatography.

12. To assess the purity of the myoglobin extract, SDS-PAGE should reveal a single protein band with a molecular weight 17-18 kDa.

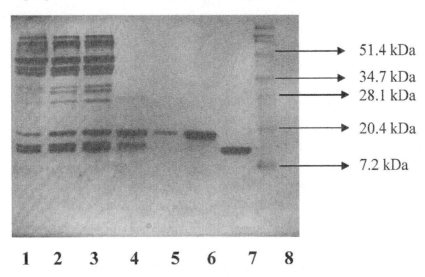

SDS-PAGE pattern of samples obtained during purification of turkey myoglobin. Lane 1, crude extract; lane 2, ammonium sulfate precipitated fraction; lane 3, dialyzed fraction; lane 4, gel filtered myoglobin (pooled from fractions 27-33) ; lane 5, ion exchange filtered myoglobin (pooled from fractions 2-9); lane 6, horse heart myoglobin; lane 7, horse cytochrome c; lane 8, molecular weight standard. (Naveena *et al.*, 2009)

6.7.6 Estimation of myoglobin, met-myoglobin and denatured myoglobin

Myoglobin was extracted from raw and cooked patties using a modified procedures of Warris (1979).Samples were blended with 5 volumes of cold 0.04M phosphate buffer at P^H 6.8 for 10 sec (raw patties) or 20 sec (cooked patties) in a homogenizer . After standing at 1°C for 1h, the mixtures were centrifuged at 3500×g at 4° for 30min. The supernatant was further clarified by filtration through Whatman no.1 filter paper. The absorbance of filtrate was measured at 525, 572 and 700 nm using a spectrophotometer. Total myoglobin (mb), met myoglobin (met mb, % of total) and percent mb denatured during cooking (PMD) were calculated using the following formulas (Trout, 1989).

1. Mb (mg/g) = $(A_{525} - A_{700})$ × 2.303 × dilution factor

    Where Mb = deoxymb + mbo_2 +Met mb

2. % Met myoglobin = $\{1.395 - [(A_{572} - A_{700}) / (A_{525} - A_{700})]\}$ × 100

3. PMD = [1 − (mb conc. After heating / mb conc before heating)] × 100

$$\text{OR}$$

$$\text{Mb denaturation (\%)}=\frac{(\text{\% mb in raw patties}-\text{\% mb in cooked patties})}{\text{\% mb in raw patties}}\times100$$

6.8 Measurement of Warner - Bratzler Shear Force Value

Tenderness measurements by objective and mechanical means are accurate and repeatable. Of all the technique, W-B shear force is most commonly used. This instrument measures the amount of force required to shear through a sample core of meat of a specific diameter. Mastication involves cutting, shearing, tearing, grinding and squeezing. Of these, this instrument simulates only shearing action. W-B shear force value shows fairly well correlation with the sensory evaluation scores.

Of the many instruments developed for measuring food texture the Warner - Bratzler shear force meter has enjoyed the greatest popularity and is being used in numerous laboratories.

Warner developed the instrument in 1927 and in 1932 Bratzler modified the instrument. The instrument now in use is motorised to ensure a constant rate of pressure and has a triangular instead of a round blade opening. A shearing speed of 9 inches per minute is used. The dynamometer dial is calibrated to allow for reading the force directly in kilograms. This devise consists of a 1-millimeter thick blade with a hole large enough to hold a cylindrical meat sample. A core of 1 cm thickness is taken with the help of a metallic mold and is placed in the opening of the blade; the blade is then led through a slit between two shear bars. Force required to shear the sample is measured on the dial scale dynamometer and expressed as Kg force / cm^2 of the meat sample.

Take a two-inch streak of longissumus muscle cut between 13th thoracic and 1st lumbar vertebrae. With help of a metallic mold take a core of 1.27 cm thick and 5-7 cm in length from middle of the muscle sample.

Determine three shear values (anterior/middle/posterior) and take average and express as Kg force / cm^2 of the meat sample.

6.9   MEASURING TEXTURE PROFILE AND SHEAR FORCE OF MEAT SAMPLES USING TEXTUROMETER

Texture is one of the important attribute affecting consumer acceptability of meat products. Meat tenderness is an important attribute of meat texture and palatability. Tenderness of meat product is determined by extent of connective tissue in meat, myofibrillar components and cooking temperature and time. Tenderness of the meat and meat products is estimated by measuring shear force value. Other texture parameters which are analyzed are chewiness, hardness, gumminess and resilience. Chewiness (N cm) is the work needed to chew a solid sample to a steady state of swallowing (hardness × cohesiveness × springiness). Hardness (N) is the maximum force required to compress the sample (peak force during the first compression cycle). Gumminess is a characteristic for semi-solid foods with a low degree of hardness and a high degree of cohesion. Resilience is how well a product "fights to regain its original height". Resilience is measured on the withdrawal of the first penetration, before the waiting period is started

Procedure for measuring texture profile of meat samples using texturometer: Texture measurements in the form of a texture profile analysis (TPA) were performed at room temperature with a texturometer (Tinius Olsen, Model H1KF, 6 Perrywood Business park, Redhill, RH1 5DZ, England). Cooked meat products were cut into 1 cm cubes and used for the analysis. They were axially compressed to 50% of the original height with a flat plunger of 20 mm in diameter (P/20) at a crosshead speed of 2 mm/s through a 2-cycle sequence. The following texture parameters were measured from force deformation curves (Bourne, 1978).

Procedure for measuring shear force using texturometer: Chilled samples were equilibrated to room temperature before texture measurement. After equilibration, 1.25 cm cores were taken using tissue borer with muscle fibres parallel to the direction of the borer. The Warner-Bratzler shear force (WBSF) of the cores were measured using Texturometer (Tinius Olsen, Model H1KF, 6 Perrywood Business park, Redhill, RH1 5DZ, England) with V-shaped stainless steel Blade (60° angle) and triangular whole in the middle. The cores were sheared perpendicular to the muscle fibre orientation with 75 Newton load range and a crosshead speed set at 200 mm/min. The force required to shear the samples was recorded in Newton (N).

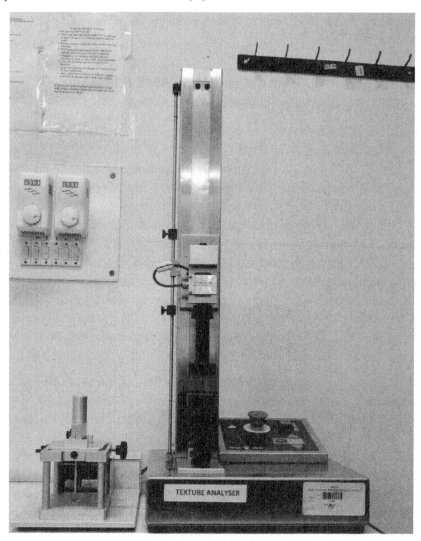

Meat texture analyzer

6.10 DIFFERENTIATION OF CHILLED AND THAWED FROZEN MEAT

This test is used to differentiate chilled and thawed frozen meat as some suppliers sell thawed frozen meat as chilled meat, since the former is likely to be of poor keeping quality and liable to spoilage and rancidity, it is important that the two be differentiated.

Two tests are available such as photometric and colour test. Both of which measure the concentration of a mitochondrial enzyme (b- hydroxy Co-A – dehydrogenase- HADH) in meat juice. The HADH activity is higher in frozen meat because freezing ruptures mitochondria, which releases this enzyme in to meat juice.

Preparation of Meat Press Juice

Press at least 10 g of intact meat piece between two glass plates under a pressure level of 10kg / cm^2. Collect the juice and store in frozen state up to one week if not analysed immediately.

Photometric Method

Reagents

0.1 M phosphate buffer, pH 6.0 - dissolve 13.6 g potassium dihydrogen orthophosphate and 14.2 g disodium hydrogen orthophosphate in 1000 ml of distilled water. Adjust the pH to 6.0 (this buffer can be stored under refrigeration for several months).

Ethylene diamine tetra acetic acid (EDTA), 34.4 mM – dissolve 10 mg EDTA in 1 ml of distilled water (this can be stored under refrigeration for several months).

β -nicotinamide adenine dinucleotide reduced form (β- NADH) grade III (7.5 mM) – dissolve 5 mg β NADH in 1 ml of distilled water (this can be stored under refrigeration for several months).

Acetoacetyl –co-enzyme A (5.9 mM)- dissolve 5 mg Acetoacetyl –co-enzyme A in 1 ml of distilled water (this can be stored under refrigeration for several months).

Procedure

Pipette 2.6 ml 0.1 M phosphate buffer, 0.2 ml EDTA, 0.05 ml NADH and 0.1 ml diluted meat juice (1:200). Start the reaction by adding 0.05 ml acetoacetyl co-enzyme. Mix and immediately read extinction at 340 nm against air. After 3 min again read the extinction at 340 nm at 25°C. Calculate the decrease in extinction per min (DE / min).

Express HADH activity as IU per ml press juice.

U/ ml = V/ E × d × n

V = volume of the test mixture (3 ml)

E = extinction coefficient for NADH at 340 nm

D = light path of cuvette (1 cm)

n = volume of meat juice (0.1 ml)

U/ml = 3 / 6.3 × 1 × 0.1 × (DE / min) × 200.

Colour Test

Reagents same as for the above stated photometry method except following solutions.

β -NADH, 1.5 mM solution (dissolve 1 mg NADH in 1 ml distilled water). This solution can be kept in refrigerator for about 4 weeks.

Meldolablue solution- Dissolve 28 mg Meldolablue in 100 ml distilled water. This solution can be kept in refrigerator for about 4 weeks.

Procedure

Pipette the following solutions into a test tube and mix.

2.4 ml 0.1 M phosphate buffer

0.2 ml EDTA (34.4 mM)

0.2 ml NADH (1.5 mM)

0.12 ml Acetoacetyl co-A (5.9 mM)

0.1 ml diluted meat juice (1:100)

Leave the test tube in dark for 60 min. then add 0.1-ml meldolablue solution and shake for 30 seconds. If the solution remains colourless the meat is fresh meat and if it turns purple the meat is frozen thawed. After 45 seconds fresh sample will go purple due to oxidation of meldolablue by air. The interpretation is based on the details given as per Table 6.1.

References

AOAC (1975)

Aragnosa, F.C. and Henrickson, R.L., (1969), Food Technol., 23: 1061-1065.

Church, P.N. and Wood, J.M. (1992) In the manual of manufacturing meat quality, Elsevier Applied Science Publishers Limited, Crown house, Linton road, Barking, Essex, IG11 8JU, England.

Table 6.1. Conditions for HADH freezing test

Species	Photometric method		Colour test	
	Dilution of meat press juice	HADH activity U/ml in frozen meat	Dilution of meat press juice	Reaction time in minutes
Beef	1:200	>3.5	1:100	60
Veal	1:200	>3.5	1:100	60
Pork	1:200	>6.0	1:100	30
Mutton	1:200	>5.0	1:100	45
Chicken (breast meat)	1:200	>5.0	1:100	60
Duck (breast meat)	1:400	>20.0	1:400	40
Goose (breast meat)	1:400	>20.0	1:400	40
Turkey (breast meat)	1:200	>5.0	1:100	60

Source: P.N. Church and J.M. Wood (1992), In the manual of manufacturing meat quality, Elsevier applied science Publishers limited, Crown house, Linton road, Barking, Essex, IG11 8JU, England.

Cross, H.R., (1979) Effects of electrical stimulation on meat tissue and muscle - A review. J.Fd. Sci, 44: 509.

Davis, G.W. Duston, T.R., Smith, G.C. and Carpenter, Z.C. (1980), Fragmentation procedure for bovine longissmus dorsi muscle as an index of cooked steak tenderness. J.Fd.Sc., 45: 880.

Faustman, C. and Phillips, A.L. (2001). Measurement of discoloration fresh meat. Ch. F3 Unit F3.3. In current Protocols in Food Analytical Chemistry; Wiley: New York, 2001.

Gornall, A.G., Bardanill, C.T. and David, M.M. (1949), Determination of serum proteins by means of the Biuret reaction, J.Biol.Chem., 177: 751.

Hamm, R. (1960) Advances in Food Research, 10: 355.

Hunter, R. S., & Harold, R. W. (1987). The measurement of appearance. New York: Wiley.

Jeremiah, L.E. and A.H. Martin, (1977), The influence of sex within breed of sires groups upon histological properties of bovine longissimus dorsi muscle during post-mortem aging. Can.J. Anim. Sci., 57: 7

Kowale, B.N., Kulkarni, V.V. and Kesava Rao, V. (2008) Methods in meat science. Pub. by Jaypee brothers medical publishers (P) Ltd., Ansari Road, Daryagunj, New Delhi.

Naveena, B.M., Faustman, Tatiyaborworntham, S., Yin, S. Ramanathan, R. and Mancini, R.A. (2009). Mass spectrometric characterization and redox instability of turkey and chicken myoglobins as induced by unsaturated aldehydes, Journal of Agircultured and Food Chemistry, 57: 8668-8676.

Oslon, D.G. and Parrish, F.C. Jr and Stromer, M.H. (1976), Myofibrillar fragmentation and shear resistance of the bovine muscles during post-mortem storage, J. Food Sci., 41: 1036.

Swift C.E., Lockett C and Fryer A.S. (1961) Food Technology, 15, 468

Swift C.E., Lockett. C and Fryer A.S. (1961) Food Technology, 15, 468

Tech, 14:75-80

Trout ,G.R.(1989).Variation in myoglobin denaturation and colour of cooked beef, pork and turkey meat as influenced by PH, Sodium chloride, Sodium tripolyphosphate, and cooking temperatures. Journal of Food science, 54: 536-544.

Tuma, H. J., Venable, J. H., Wuthier, P. R., &Henrickson, R. L., (1962). Relationship of fibre diameter to tenderness and meatiness as influenced by bovine age. *Journal ofAnimal Science, 21,* 33-36.

Wardlow, F.B. Mc Caskill, L.H. and Acton, J.C. J., (1973) Food Science, 38: 421

Warris, P.D. (1979).The extraction of haen pigments from fresh meat. J.Food

Whiting, R.C. and Jenkins R.K., (1981) J. Food Sci., 46: 1693-1696.

CHAPTER-7

■ ■ ■

ANALYSIS OF CURING INGREDIENTS IN MEAT AND MEAT PRODUCTS

7.1 ESTIMATION OF SALT CONTENT

Salt is the basic of all the curing mixtures and is the only ingredient necessary for curing. Salt act by dehydration and altering of the osmotic pressure so that it inhibits bacterial growth and subsequent spoilage. Use of salt alone, however, gives a harsh, dry, salty product, that is not palatable. Salt alone results in a dark undesirable coloured lean that is unattractive and objectionable to consumers. As a consequence salt is generally used in combination with both sugar and nitrate/ nitrite. The level of salt used in meat products is generally up to the taste or 2 to3 percent; while pickling cure strength of 50 to 85° (salinometer strength) is most common.

Principle

The sample is digested with nitric acid in the presence of an excess of silver nitrate. The excess silver nitrate is then titrated with ammonium thiocyanate and the salt calculated from the amount of Ag + required to precipitate the Cl ⁻ present in sample.

Apparatus

1. Hot plate,
2. Fine Balance,
3. Burettes— 25 ml and / or 50 ml capacity,
4. Acid dispensing burette with stopcock,

5. Erlenmeyer flasks, pyrex, wide mouth, 250 ml capacity,
6. Beakers — 50 ml or 100 ml,
7. Graduated cylinders — 25, 50 and 100 ml,
8. Pipettes — 5,10 and 25 ml,
9. Volumetric flask — 1000 ml capacity.

Reagent

1. Sodium chloride,
2. Silver nitrate,
3. Ammonium thiocyanate,
4. Ferric ammonium sulphate,
5. Nitric acid free from chlorides, specific gravity 1.42,
6. Potassium permanganate,
7. Diethyl ether.

Solutions

1. Silver nitrate (MW 168.89) solution (0.50 N) accurately standardised against sodium chloride solution containing known amount of pure dry NaCl. Weight 168.89 g of Silver nitrate and dilute in volumetric flask with distilled water, making 1 liter of 1.0 N or weigh 84.945 g and dilute to 1 liter in same volume making 0.5 N of AgNO3.
2. Ammonium thiocyanate (M.W. 76.13) solution (0.10 N), accurately standardised against silver nitrate solution. Weigh 76.13 g of NH_4SCN and dilute in water and make up to 1 liter to make 0.10 N solution.
3. Potassium permanganate 10 percent in water.
4. Ferric ammonium sulphate indicator solution, saturated (135 g per 100 ml of water). 100 ml solution is acidified with 5 ml of nitric acid and prepared a day before use to make sure that all the salt is dissolved at room temperature.

Procedure

1. Weight 2.5, 5.0 or 10.0 g of sample and place in 250 ml Erlenmeyer flask.
2. Add 5 or 10 ml of 0.5 N silver nitrate, depending on salt content. Sample must be thoroughly moistened with added silver nitrate.
3. Add 15 to20 ml of nitric acid (specific gravity 1.42).

4. Place on heater in hood and allow solution to come to a boiling point. Boil for a couple of minutes until sample is dissolved.

5. Add 10 percent concentrated potassium permanganate solution in small portions and keep boiling until permanganate colour disappears and solution becomes colourless or nearly so.

6. Add about 25 ml of distilled water.

    a. Boil for another 5 minutes.

    b. Cool and add water making to about 150 ml

    c. Add 5 ml of ferric indicator

10. Add about 10 ml of diethyl ether

11. Titrate with 0.1 N NH_4SCN.

Calculate as follows

$$\text{Percent NaCl} = \frac{(A - 0.2\ B) \times 2.92}{\text{Weight of sample}}$$

A = ml 0.5 N $AgNO_3$ added to sample

B = ml 0.1 N NH_4SCN used to titrate sample

Clarification of Formula

1 ml of 0.1 N $AgNO_3$ reduces 0.0584 percent of NaCl.

1 ml of 0.5 N $AgNO_3$ reduces 0.292 percent of NaCl.

10 ml of 0.5 N $AgNO_3$ reduces 2.92 percent of NaCl.

So if $AgNO_3$ of 0.5 N is placed in amount of 11 ml and back titration is 5 ml 0.1 N, thus amount used for 10 g of sample is 1 ml of 0.5 N $AgNO_3$.

$$\text{Salt} = \frac{(11 - 1) \times 2.92}{10} = 2.92 \text{ percent}$$

7.2 ESTIMATION OF CHLORIDE CONTENT

Apparatus and reagents

1. Volumetric flask (200 ml)

2. Conical flask (250 ml)

3. Burette, pipette

4. pH meter

5. Nitrobenzene

6. Nitric acid solution about 4 N (mix 1 volumes of concentrated Nitric acid with 3 volumes of distilled water)

7. Solution for precipitation of proteins

   a) Reagent I – dissolve 106 g of potassium ferricyanide ($K_4Fe(CN)6.3$ H_2O) in water and dilute to 1000 ml.

   b) Reagent II – dissolve 220 g of zinc acetate (Zn ($CH_3COO).2H_2O$) and 30 ml of glacial acetic acid in water and dilute to 1000 ml.

8. Standard N /10 silver nitrate solution (dry $AgNO_3$ for 20 hours at 150°C and allow to cool in a desiccator. Dissolve 16.989 g of dried salt in water and dilute to 1000 ml.

9. Standard N /10 potassium thiocynate solution – dissolve about 9.7 g of potassium thiocynate (KCNS) in water and dilute to 1000 ml. Standardise the solution to nearest 0.0001 N against silver nitrate.

10. 1 N NaOH solution

11. Ammonium iron sulphate ($NH_4Fe(SO_4)2.12 H_2O$) saturated solution in water

12. Activated charcoal.

Procedure

Weigh 10 g meat sample and transfer into conical flask. Add 0.5 g of activated charcoal and 100 ml of hot water. Heat for 15 min in boiling water bath. Cool and add 2 ml of reagent I and II each. Adjust pH to 7.5-8.3 using NaOH. Allow the flask to stand for 30 min. transfer the contents to 200 ml volumetric flask and make the volume. Mix and filter through fluted filter paper. Transfer 20 ml of filtrate to conical flask and add 5 ml HNO_3 solution and 1 ml of ferric ammonium sulphate indicator. Add 20 ml of $AgNO_3$ and 3 ml of nitrobenzene. Mix, shake vigorously to coagulate proteins. Titrate the supernatant with potassium thiocynate.

Chloride content as percent sodium chloride = 5.844 (20-V) / Weight of sample

   V = ml of potassium thiocynate required for titration

7.3 ESTIMATION OF NITRITE CONTENT IN MEAT BY SULFANILIC ACID METHOD

The nitrate / nitrites are added in meat products to stabilize meat colour of lean tissue, impart flavour to cured meats, to inhibit growth of many food poisoning bacteria and especially Clostridium botulinum and to retard the development of rancidity. The level of nitrate or nitrite used in meat is 500 and 250 ppm respectively. Nitrosamines, a carcinogenic compound are formed from nitrites during heating of meat products. Use of Sodium ascorbate (500ppm) in combination with Sodium nitrite (120ppm) is effective in reducing the nitrosamine formation while keeping the beneficial effects of nitrites intact.

Principle

Extracted aqueous sample from meat is subjected to react with GRIESS reagent, containing sulfanilic acid, alpha-naphthylamine HCl. Developed colour is measured spectrophotometrically and the nitrate content is determined by a reference to a standard curve, prepared by measuring the colour developed in solution containing a known amount of nitrate.

Apparatus

1. Balance - Analytical
2. Flasks: volumetric, glass stoppered, 50, 100, 500 and 1000 ml
3. Pipettes: volumetric, 1,2,5,10 and 50 ml
4. Pipettes: graduated for 10 ml in 0.1 ml divisions
5. Beakers: 50 and 100 ml
6. Graduated cylinders: 10,100 and 250 ml
7. Flasks: Erlenmeyer, 225, 250 and 500 ml
8. Funnels: 75 mm., 3 inch in diameter
9. Filter paper: Whatman 42 or equivalent, 12.5 cm diameter
10. Steam bath.
11. Spectrophotometer
12. Cuvette, selected to match transmittance at 520 nm (1 and 2 cm in diameter).

Reagents

1. Glacial Acetic Acid, 99.5percent
2. Sulfanilic Acid
3. Alfa-naphthylamine Hydrochloride,
4. Sodium Nitrate,
5. Hydrochloric Acid, specific gravity 1.18
6. Mercuric Chloride.

Solutions

Modified Griess reagent

1. Sulfanilic acid solution: Dissolve 0.5 g of sulfanilic acid in 150 ml of 15 percent acetic acid (V/V). (22 ml glacial acetic acid and 128 ml of distilled water)

2. Alfa-naphthylamine HCl solution: Boil 0.1 g of this chemical in 20 ml of water until dissolved and pour while hot into 150 ml of 15 percent acetic acid.

3. Mix solution (1) and solution (2) and store in tightly closed brown glass bottle. This is a modified Griess reagent. Keep in refrigerator.

4. Mercuric chloride solution, saturated (about 70 g in 1 liter of water at 20°C.)

5. Standard solution of sodium nitrite (5 mg /ml):Weigh exactly 1 g sodium nitrite, (with assay of 97 percent) and transfer to a 1000 ml volumetric flask, make up volume with distilled water and mix thoroughly. Pipette 50 ml of this solution into another 1000 ml volumetric flask, dilute to volume and mix thoroughly again. Pipette 50 ml of this dilute solution into 500 ml volumetric flask, fill up with water to volume and mix thoroughly. One ml of this solution contains 0.005 mg of sodium nitrite, i.e. 5 mg.

Standard Curve

1. With a pipette transfer into 50 ml volumetric flasks zero (Blank) 1.0 ml, 2.0 ml, and so on up to 10 ml of the dilute standard nitrite solution. Add water to about 30ml in each flask and mix.

2. Add 2 ml of Griess reagent and make up volume with distilled water. Mix thoroughly and allow to stand at room temperature for 1 hour.

3. Read at 520 nm.

4. Read optical density of each standard solution from 1 ml up to 10ml of prepared nitrite solution with 5 mg in each 1 ml

5. Plot a graph of the optical density readings against the nitrite concentration.

Procedure

1. Weigh 5.0 g of prepared meat sample in a beaker and place in 250 ml volumetric flask. Add about 100 ml of water and break the sample with stirring rod.

2. Place in water bath and heat to 80°C. Keep mixing the sample and break up all lumps that may form.

3. . At the end of heating in water bath, pipette into each sample 5 ml of saturated Mercuric chloride ($HgCl_2$) solution.

4. Cool and make volume to 250 ml mark.

5. Filter through 12.5 cm filter paper and transfer 10 ml of filtrate to a 50 ml volumetric flask. This will represent 1/5 g of sample in 50 ml flask.

6. Add water to about 30 ml, mix and add into 2 ml of Griess reagent, make volume to 50 ml with water and mix.

7. Allow the colour to develop for 40 min to 1 hr at room temperature in diffuse light.

8. Determine the optical density at a wavelength of 520 nm with the adjusted zero optical density with blank, containing 2 ml of reagent in 50 ml water solution.

9. Read the Sodium nitrite content of the solution from the standard curve.

    (AOAC, 1975)

7.4 ESTIMATION OF NITRITE IN CURE AND PICKLE (IN THE PRESENCE OF ASCORBATES)

Principle

Method used for cures and pickles is similar to that of meats but time for diazotization (3 minutes) and for coupling (3 minutes) enables better recovery by employing different reagents.

Reagents

1. Sulfanilamide solution- Dissolve 0.5 g in 100 ml of 1:1 hydrochloric acid.

2. N- (1- napthyl) ethylenediamine dihydrochloride solution - dissolve 0.1 g in 100 ml of distilled water.

Procedure

1. Prepare a proper dilution for cures and pickles, which depends on suggested concentration of Sodium nitrite.

2. To the 10 ml of cure or pickle diluted solution transferred into 100 ml of volumetric flask, add: a)2.0 ml of sulfanilamide solution swirl and allow to stand for 3 minutes, then add b) 2.0 ml of N-(1-napthyl) ethylenediamine

dihydrochloride solution , dilute to 100 ml with distilled water, mix well and let stand for another 3 minutes before final reading.

3. Measure optical density at 540 nm, setting at 100percent level with a reagent - blank (2 ml of each solution + 98 ml water).

4. Obtain sodium nitrite concentration from standard curve.

Suggested Solutions

1. Pickle liquid: Take 25 g of pickle liquid and dilute to 250 ml by distilled water in volumetric flask. 1 ml this solution will contain 0.1 g of pickle.

2. Dry cure mix: Dilute 5 g in 500 ml, volumetric flask and bring to mark with distilled water. 1 ml of solution will contain 0.01 g of cure.

3. Standard curve may be prepared separately for dry cures that require different dilutions and for pickles with less concentrated nitrite. Standard curve for cure is suggested to be prepared on 0.001 g / ml basis.

7.5 ESTIMATION OF NITRATES IN MEAT (XYLENOL METHOD)

Principle

Nitrates are extracted from the meat sample with hot water. Nitrites are oxidised to nitrates. m - Xylenol is added and the coloured complex is distilled into dilute alkali. The colour is measured by spectrophotometer and nitrates are determined by a standard curve. (When nitrates are present appropriate correction should be made in final calculation.)

Reagents

1. Meta-Xylenol, 1 hydroxy-2, 4 dimethyl benzene
2. Silver sulphate, nitrate free
3. Potassium permanganate, crystals
4. Ammonium hydroxide
5. Phosphotungstic acid, crystals, Analytical reagent
6. Sulfuric acid, low nitrogen, 98.9percent
7. Sodium hydroxide, pellets
8. Bromocresol green
9. Sodium nitrate.

Solution

1. Bromocresol green indicator. Dissolve 0.1 g of Bromocresol green in 1.5 ml 0.1 N sodium hydroxide and make to 100 ml with distilled water.

2. Sulfuric acid solution, 1 part (volume) of H_2SO_4 plus 10 parts of water.

3. Sulfuric acid solution. 3 parts (volume) of H_2SO_4 plus 1 part of water.

4. Silver ammonium hydroxide solution. Dissolve 5 g of silver sulphate in 60 ml NH_4OH solution. Heat to boiling, concentrate to 30 ml, cool and dilute to 100 ml with distilled water.

5. Potassium permanganate. Take 25 g of pickle liquid and dilute to 250 ml by distilled water in volumetric flask. 1 ml this solution will contain 0.1 g of pickle. 0.2N solution.

6. Phosphotungstic acid solution. 20 g per 100 ml

7. Sodium hydroxide solution 10 g NaOH per liter.

8. Sodium nitrate solution. Dissolve 0.2 g of sodium nitrate in water and make to 1 liter.

Procedure

1. Weigh accurately 10 g of prepared sample into 250 ml beaker Mix thoroughly with 80 ml warm water and heat on steam bath for 1 hour stirring occasionally.

2. Cool to room temperature and make to 100 ml.

3. Mix. Filter and transfer 25 ml of filtrate to a 50 ml volumetric flask.

4. Add three drops of Bromocresol green indicator and dilute H_2SO_4
   (1:10 as indicated in Sol. 2) drop wise with shaking until colour changes to yellow.

5. To oxidise nitrites to nitrate add 0.2 N $KMnO_4$ drop wise with shaking until faint pink colour remains for 1 minute.

6. Add H_2SO_4 (3:1 as indicated in Sol 3) and phosphotungstic acid solution (Sol 6) also 1 ml each in order to precipitate proteins. Make to volume (50 ml). Mix and filter.

7. Transfer a suitable aliquot not more than 20 ml (containing 0.2 to 2.0 mg $NaNO_3$)to a 500 ml Erlenmeyer flask and add sufficient silver ammonium hydroxide (2ml) to precipitate all chlorides and excess of phosphotungstic acid.

8. Add a volume of H_2SO_4 (3:1) three times the volume of the liquid in flask. Stopper, mix and cool to 35°C. Add 1 to 2 drops of meta-xylenol, stopper and

shake. Yellow to brownish yellow colour indicative of nitrates appears, bright red precipitate may also appear due to excess of phosphotungstic acid.

9. Let stand for 30 minutes at 30-40°C, add 150 ml of water and distill into alkali containing 5 ml of NaOH in graduated cylinder (Sol. 7).

10. Collect 50 ml of distillate, transfer to 100 volume flask and fill up with water to the mark. Mix well.

11. Transfer solution to the spectrophotometer and determine optical density at 450 nm using water as blank.

12. Determine nitrate in the solution from prepared standard curve;

$$\text{Percent Sodium nitrate} = \frac{20A}{B \times C} \text{ minus } 1.232 \text{ D}.$$

A = NaNO$_3$ content (mg) in the solution as read from the standard curve.

B = Sample weight in g.

C = Aliquot (ml) used for nitration (See pt. 7).

D = Percent sodium nitrite determined.

Standard Curve for Nitrate

1. Transfer 0, 1.0 ml, 2.0 ml and so on up to 10 ml of the standard nitrate solution (0.2 g made to 1 liter) into a series of 500 ml Erlenmeyer flasks and add 10 ml distilled water in each flask.

2. Add 30 ml H$_2$SO$_4$ (3:1) to each flask mix, cool to about 35°C, add 1-2 drops on metaxylenol, stopper, shake and hold at 30-40°C for 30 minutes.

3. Add 150 ml water, taking care to wash off the stopper.

4. Distill 40-50 ml from each flask, using Kjeldahl assembly, into 100 ml graduated cylinder containing 5 ml NaOH (Sol.7).

5. Transfer the distillate from each distillation to separate 100 ml volumetric flasks. Make each to the volume with distilled water and mix.

6. Read the optical density at 450 nm and prepare the graph.

(AOAC, 1975)

7.6 ESTIMATION OF SUGAR CONTENT

7.6.1 Total Sugars in Meat Products, Cure and Pickle

The addition of sugars in cures is primarily for flavour. Sugar softens the product by counteracting the harsh hardening effect of salt, by preventing some of the moisture removal and by a direct moderating action on flavour. Sugars also interacts with the amino groups of the protein and when cooked forms browning product that enhances the flavour of cured meat. Sugars are added at a level of 2 percent in cure mix.

Principle

Total carbohydrates determined before and after hydrolysis. Copper is estimated quantitatively and calculated as sugars.

Apparatus

1.   Balance, analytical
2.   Waring blender
3.   Pipettes, measuring, 10-25 ml
4.   Water bath water
5.   Water bath for cold water (to cool samples)
6.   Erlenmeyer flasks, wide mouth, 225-250-500 ml
7.   Graduated cylinders, 50, 100, 250, 500, 1000 ml
8.   Stoppers, some with hole to fit bottles and flasks.
9.   Whatman filter paper, rapid, moderately retentive No.54
10. Whatman filter paper No. 42, very retentive and strong.
11. Funnels, average stems, analytical, polypropylene
12. Hot plate
13. Litmus paper-acid, alkali
14. Thermometer
15. Volumetric flasks:2-500 ml, 1-1000 ml capacity

Reagents

1.   **Modified Fehling solution**
     a.   Copper sulphate: dissolve 34.639 g of copper sulphate in distilled water and dilute to 500 ml.
     b.   Potassium sodium tartrate (Rochelle salt): Weigh 173 g and add to dissolved 50 g of sodium hydroxide (pellets) in distilled water and bring to 500 ml.

     Reagents should be stored in dark with tightly stoppered bottles.

2. Phosphotungstic acid, 20percent solution in water.

3. Hydrochloric acid (specific gravity1.19)

4. Sodium hydroxide, pellets

5. Methyl red indicator 0.1 percent in alcohol

6. Ferric sulphate solution, dissolve 55 g of anhydrous ferric sulphate and make with distilled water to 1000 ml.

7. Sulfuric acid 4.0 N

8. Potassium permanganate standard solution. Prepare 0.1573 N solution that contains 4.98 g in 1000 ml of distilled water.

7.6.2 Estimation of Invert sugar in Meat Products

1. Weigh 10 g of sample.

2. Add about 200 ml of cold distilled water and 20 ml of copper sulphate solution and blend in Waring Blender for about 2 minutes and transfer to 250 ml volumetric flask.

3. Place in hot water bath (82°C) and occasionally mix with stirring rod. Keep in water bath for about 15-20 minutes.

4. Cool the sample to room temperature in cold water bath and fill to the mark of 250 ml with distilled water.

5. Prepare another flask with wide mouth-Erlenmeyer type. Place the funnel and Whatman filter paper No 54 and filter 100 ml of filtrate.

6. Add to the filtrate 10 ml of concentrated hydrochloric acid, and place again in hot water bath for hydrolysis. Keep warming up until sample reaches about 75°C. Maintain this temperature for another 10 minutes.

7. Cool sample in cold water bath until reaches room temperature and add 9.5 ml of methyl red indicator.

8. Swirl flask in rotation way and add enough of saturated sodium hydroxide (about 6.0 ml) to bring the pH to about slightly acidic state.

9. Place in cold water bath again and add about 30 ml of 20 percent phosphotungstic acid solution in order to precipitate proteins.

10. Filter entire sample, using Whatman filter paper No. 54, into 500ml Erlenmeyer flask. This will represent 4 g of meat product sample.

11. Add 35ml of Rochelle salt solution, and 20 ml copper sulphate solution. Place the flask on hot plate and bring to boil. Use stopper with hole in order to bring to boiling in about 4 minutes. Boil for another 2 minutes when copper

should precipitate in the flask. If not add more Rochelle salt solution up to 10 ml and bring to boiling.

12. Filter the precipitate while still hot using Whatman No. 42. At the end of filtration add some portions of distilled water at room temperature onto filter paper for flushing until you reach almost neutral drip of water from filter paper.

13. Place the filter paper with residual copper into separate 500 ml flask. Add 200 ml of distilled cold water, and add about 50 ml of ferric sulphate solution.

14. Tear off filter paper with stirring rod and keep mixing until all copper residue is dissolved and colour becomes clear brownish.

15. Add 25 ml of 4.0 N sulfuric acid and shake until green colour appears.

16. Titrate the contents of a flask with 0.1573 N potassium permanganate until purplish colour appears and persists for another 30 seconds.

17. Deduct 0.2 to 0.3 ml for blank. Calculate on the basis of on standard sugar solution titration, or use Hammond's Table in A.O.A.C. methods of analysis. Divide results by number of g used for analysis and express in percentage as invert sugar.

7.6.3 Estimation of Invert Sugar in Cure and Pickle

For cure containing sugar use about 0.1 g of cure. Dilute 2.5 g in 250 ml of water, and use 10 ml of this solution. Add 90 ml of water and bring to 100 ml representing 0.1 g of cure.

For pickle solution. Weigh 25 g, dilute to 250 ml, and use 10 ml of solution, add 90 ml of water. This will represent 1 g of pickle solution. Once sampling is completed (cure 100 ml =0.1 g, pickle 100 ml = 1.0 g) proceed as follows:

1. Add 10 ml of concentrated hydrochloric acid and place in hot water bath for hydrolysis (inversion). Keep warming up until reaches 82°C. Maintain this temperature for another 10 minutes.

2. Cool sample in cold water bath to room temperature and add 0.5 ml of methyl red indicator.

3. Swirl flask and add enough of saturated sodium hydroxide (about 6.0ml) to bring the pH to about slightly acidic state.

4. Add about 5 ml of phosphotungstic acid solution to precipitate possible protein present in pickle or cure.

5. Filter sample (Whatman No. 42) into 500 ml of Erlenmeyer flask.

6. Add Rochelle salt solution about 35 ml and copper sulphate solution 20 ml sample.

7. Place flask on hot plate and bring to boil in about 4 minutes. Keep boiling for 2 minutes. Copper should precipitate this time.

8. Filter the precipitate while still hot through filter paper, Whatman No.42 and by the end of filtration and wash the residue with distilled water until neutral drip of water is obtained.

9. Transfer filter paper with residual copper to another flask. Add 200 ml of distilled water, 50 ml of ferric sulphate solution and tear filter paper completely .Be sure that copper residue is dissolved.

10. Add 25 ml of 4.0 N sulfuric acid and shake until colour becomes green.

11. Titrate the contents with 0.1573 N potassium permanganate until purplish colour appears and persists for another 30 seconds.

12. Read results from Hammond Table in A.O.A.C. methods of analysis and express as invert sugar, which may be calculated for sucrose level if sucrose is added (See table)

Excerpts from the Hammond table for calculating dextrose (glucose) lactose, invert sugar and sucrose (expressed as mg.)

Copper (Cu)	Glucose (Dextrose)	Lactose H_2O	Invert Sugar	Invert Sugar & Sucrose (g of total sugar)		
				0.3 g	0.4g	2.0g
10	4.6	7.7	5.2	3.2	2.9	--
20	9.4	15.4	10.2	8.3	7.9	1.9
30	14.3	23.0	15.3	13.4	13.0	7.0
40	19.2	30.6	20.4	18.6	18.2	12.1
50	24.1	38.3	25.5	23.8	23.3	17.3
60	29.0	46.0	30.6	28.9	28.5	22.5
70	34.0	53.6	35.8	34.2	33.7	27.7
80	39.0	61.3	41.0	39.4	38.9	32.9
90	44.0	69.0	46.2	44.7	44.2	38.2
100	49.0	76.7	51.5	50.0	49.5	43.5
150	74.7	115.2	78.1	76.8	76.3	70.4
200	101.1	153.9	105.4	104.3	103.8	98.0
250	128.2	192.9	133.4	132.5	132.0	126.5
300	156.2	232.0	162.2	161.4	161.0	155.7
350	185.0	271.3	191.8	191.0	190.8	185.9
400	214.7	311.0	222.2	221.5	221.5	217.0

7.7 ESTIMATION OF ASCORBIC ACID CONTENT

7.7.1 Estimation of Ascorbic Acid Content in Meat Products

Principle

The property of reduction of the colour of 2,6- dichlorophenolindophenol dye by ascorbic acid is used this method.

2,6- dichlorophenolindophenol dye is reduced equally by ascorbic acid, sodium ascorbate, erythorbic acid and sodium erythorbate.

Reagents

1. Dye solution: Dissolve 130 mg of the Sodium salt of 2.6 - dichlorophenolindophenol in approximately 250 ml of hot water containing 210 mg of $NaHCO_3$ (sodium bicarbonate), cool, filter and dilute to one liter. Place in a brown bottle and store in refrigerator. Standardize the solution before use.

2. Phosphate Buffer (pH 6.0): Mix 1 volume of 10percent (W/V) Na_2HPO_4 (sodium phosphate, dibasic) with 5 volumes of 10percent (W/V) $NaH_2PO_4H_2O$ (sodium phosphate monobasic). This buffer solution has a pH of approximately 5.6, but when it is diluted as in the procedure (e.g. 1 volume with 10 or more volumes of water) the pH is about 6.0.

3. Standard solution: Dissolve 50 mg ascorbic acid in water and dilute to 250 ml. To a 5 ml aliquot of this solution, containing 1 mg ascorbic acid, add 2 ml of the phosphate buffer about 50 ml of water and titrate with the dye solution to a stable blue colour. The ascorbic acid equivalent of 1 ml of dye should be calculated in ppm.

Apparatus

1. Waring Blender
2. Erlenmeyer flasks, 250 ml
3. Pipettes: 2 ml and 10 ml graduated.

Procedure

Weigh 10 g of meat sample, add sample, 200 ml of distilled water and blend for 1 minute in Waring blender. Transfer blended sample to 250 ml volumetric flask and fill up to mark with water and filter.

Transfer 25 ml to Erlenmeyer flask, add 2 ml of phosphate buffer and about 50 ml of distilled water and titrate with dye solution until light blue colour appears (persistent).

Check the strength of the dye, titrating against the standard solution and make final calculation and express in ppm.

$$\frac{10 \text{ g} \times 25 \text{ ml aliquot}}{250 \text{ ml}} = 1 \text{ gram of sample tested.}$$

In order to have a more precise test made for meat products containing sodium ascorbate or equivalent, check same procedure with 10 g samples of untreated meat. Titrated dye to a persistent colour will be considered as blank, which should be deducted from the total titration of tested meat samples.

7.7.2 Estimation of Ascorbate Contents in Pickles and Cure

Principle and reagents for tests are similar to those shown in the method for "Ascorbates in Meat". For pickles and cure, however, it is better to use buffered salt solution, which prevents the sample against breakdown when pH is raised over 6.

To 1000 ml of distilled water add 150 g of sodium chloride (salt) and 5 ml of phosphate buffer. This will represent buffered 15percent salt solution.

1. **Cure:** Weigh 2.5 g accurately and transfer to 250 ml volumetric flask. Shake well with added buffered salt solution, fill up to the mark. Pipette 10 ml and transfer to 100 ml volumetric flask, fill up with same salt solution to mark and again pipette 10 ml to another flask. This will represent 0.01 g of sample.

$$\frac{2.5 \times 10 \times 10}{250 \times 100 \times 10} = 0.01 \text{ g of sample}$$

Add 50 ml of same salt solution, add few drops of phosphate buffer and titrate with dye solution. There should be no need to compare with blank sample, since 1 drop of dye will make solution blue. If reducing capacity of dye is calculated on 1 g of sample, multiply the results by 100.

2. **Pickle:** Procedure similar except that weigh 25 g of pickle, dilute with buffered salt solution into 250 ml flask and then pipette 10 ml of the solution to the flask. Add again salt solution, buffer and titrate. Sample will represent 1 g of pickle:

$$\frac{2.5 \times 10}{250} = 1 \text{ g}$$

If it is too concentrated, repeat the test with more diluted sample.

7.8 ESTIMATION OF PHOSPHATE CONTENT IN MEAT PRODUCTS AND CURING PICKLES

Phosphates when added to the cure increase the water binding capacity and thereby the yield of the finished product. The action of phosphate in improving water binding capacity is two fold: raising the pH and causing unfolding of the muscle proteins, thereby making sites available for water binding. Only alkaline phosphates are effective for improving water binding since acid phosphates may lower the pH and cause greater shrinkage. Phosphates also chelates trace metal ions and retard the development of rancidity in meat products. Because of corrosive nature of phosphates the equipment utilised must be made of stainless steel or plastic. Sodium tripolyphosphate, Sodium hexametaphosphate, Sodium pyrophosphate and disodium phosphate have been approved for adding in cure mix. The legal limit for added residual phosphate is set at 0.5 percent in finished products. Since meat contains 0.01 percent natural phosphate, this must be subtracted in calculating the level of added phosphate during curing.

Principle

Meat sample is ashed in muffle furnace and then treated with citro-molybdate solution and finally titrated with sodium hydroxide. Consumed NaOH is calculated as percentage of P_2O_5 (Phosphorus pentoxide).

Apparatus

1. Muffle furnace
2. Oven with air convection
3. Hot plate and hot water bath
4. Porcelain dishes for ashing, dia. 80 mm, and height 30 mm.
5. Filter paper: Whatman No.54, 42
6. Erlenmeyer flask: 250, 500 ml
7. Volumetric flask : 500 and 250 ml
8. Graduated flask: 500 and 250 ml
9. Measuring pipettes
10. Balance.

Reagents and Solutions

1. Citro- molybdate solution
   a) In 1360 ml distilled water dissolve:108 g of ammonium nitrate, 105 g of citric acid, powder and 136 g of ammonium molybdate.
   b) Into 310 ml of distilled water pour 250-ml of nitric acid, (69 - 71 percent).
   c) Pour solution (a) into solution (b) and mix well.
2. Hydrochloric acid solution (1:3): HCl 1 volume + H_2O 3 volume.
3. Ammonium nitrate, reagent grade.
4. Sodium hydroxide solution made from NaOH pellets, 1.0 N, accurately standardised.
5. Sulfuric acid, (A.R.), 0.5 N, accurately standardised.
6. Phenolphthalein indicator, make 1percent solution in 95 percent alcohol.

Procedure for Meat and Meat Products

1. Weigh 10 g of sample and place in a porcelain dish for ashing.
2. Pre-ash the sample in hot air convection oven at 145°C until water is removed (vaporised).
3. Place the sample in muffle furnace and ash at 525°C for about 2 hours. Observe instructions for ashing.
4. Cool the sample and add 40 ml of diluted (1:3) hydrochloric acid.
5. Heat the sample in same porcelain dish on hot plate until ash is dissolved.
6. Filter the contents through Whatman filter paper, grade 54 and flush the crucible several times with hot distilled water, pouring the contents to same filter paper in order to remove all soluble contents. The filtrate should be in about 150 to 200 ml.
7. Add 1 g of ammonium nitrate granules, swirl and place the flask in hot water bath until 80-85°C. is reached.
8. Pour 60 ml of citro-molybdate solution and heat again in hot water bath to about boiling temperature (90-95°C.) Keep for at least 30 minutes.
9. Filter the precipitate (yellow) through filter paper (Whatman no. 42) and flush the flask with hot water several times, pouring into the filter paper until about neutral pH of dripped water from filter is achieved. (pH 6.0).
10. Place the filter paper in a 250 ml Erlenmeyer flask and add about 100 to125 ml of cold distilled water.

11. Add 1.0 N Sodium hydroxide solution mix the contents well, break up with glass rod filter paper until yellow colour disappears.

12. Add several drops of phenolphthalein indicator.

13. Shake well and titrate back with 0.5 N Sulphuric acid until red colour becomes very slightly pink. End point.

14. Calculate phosphate content as follows:

P_2O_5 percent = (NaOH consumed) × 0.309 / weight of sample

NaOH consumed = ml 1.0 N of NaOH minus back titrate $\dfrac{H_2SO_4 \ (0.5N)}{2}$

Converting Factors for Phosphates

Percent P_2O_5 expresses percent of phosphorus pentoxide. Multiplier 0.31 (or 0.309) is used because of titrating NaOH 1.0 N instead of 0.3240 N.

Percent P_2O_5 x 0.436 = percent phosphorus (P)

Percent P_2O_5 x 2.185 = percent Bone phosphate of lime (BPL)

Percent P_2O_5 x 1.74 = percent tripolyphosphate, sodium (used for hams)

Percent P_2O_5 x 1.48 = percent hexametaphosphate, sodium

Calculation for Added Phosphate

Example:

Total protein in sample analysed ..... 16.30 percent

Natural phosphate as P_2O_5 : X)

(Protein × 0.025) .... 0.41 percent

Phosphate by analysis as P_2O_5 .... 0.69 percent

Added phosphate as P_2O_5 .... 0.28 percent

Calculated added phosphate as

Tripolyphosphate (0.28 × 1.74) .... 0.49 percent

Maximum level is 0.50percent of added P_2O_5 X) P_2O_5 in protein calculated from table of natural phosphorus (P) which equals 0.436 x P_2O_5 as explained in converting factors. Natural phosphate varies in different products, depending on protein content.

Procedure for Curing Pickles

1. Weigh 25 g of pickle and dilute with water in volumetric flask to 250 ml filter if necessary.
2. Pipette 10 ml of this pickle solution that will represent 1 g of pickle.
3. Transfer 10 ml of pickle solution to separate flask, add 90 ml of water and 5 ml of concentrated nitric acid.
4. Place in hot boiling water bath and keep covered for 2 hours for hydrolysis.
5. Continue as indicated in step number 7 to 14 in the procedure for meat products.
6. Final calculation will be completed when established NaOH (1.0N) consumed will be multiplied by 0.309 considering 1 g sample used for analysis. Express results in P_2O_5 percent and convert to appropriate factors explained above.

7.9 ESTIMATION OF SULFITE CONTENT IN MEATS

7.9.1 Qualitative Estimation of Sulfite Content in Meats

Principle

Malachite green dye solution mixed up with meat changes colour and dye is decolourized when sulfites are present in meat.

Reagent

Malachite green solution: Dissolve 10 mg of malachite green in distilled water and dilute to 50 ml. Discard when visible deterioration occurs.

Procedure

1. Transfer about 3.5 g of the ground meat product to 4 / 4" square white waxed freezer paper.
2. Run a blank on a sample that is known to be free of sulfite.
3. Add 0.5 ml of malachite green solution to the ground sample, mixing vigorously for two minutes with a spatula.
4. Observe colour changes. Dye is decolourised (disappears) in the presence of sulfites. Normal meat becomes blue-green.

7.9.2 Quantitative Estimation of Sulfites in Meats

Principle

Lead acetate paper will turn dark if meat sample contains sulfite.

Standard tests made for known amount of sulfite permits to establish by comparison content of sulfite in ppm.

Apparatus

1. Erlenmeyer flask : 250 ml
2. Whatman filter paper No.1, Size A - 4.25-cm in diameter
3. Gutzeit tube, scrubber
4. Stoppers to fit Gutzeit tube
5. Analytical balance

Procedure

1. Place 25 g of ground meat into 250 ml Erlenmeyer flask. Add 50 ml of 20 percent HCl, place the stopper and mix.
2. Prepare a strip of lead acetate paper by dampening a piece of Whatman filter paper No.1 with 10percent solution of lead acetate. Blot the prepared lead acetate paper dry between pieces of filter paper. Place lead acetate paper in wide end of a Gutzeit scrubber tube fitted with new rubber stopper.
3. Remove the stopper from the 250 ml Erlenmeyer flask containing ground meat and acid. Add 5 g of granular zinc (20 mesh) and immediately insert the stopper end of the Gutzeit scrubber tube into the test flask. Gently swirl the mixture to avoid foaming.
4. The lead acetate paper will turn black within a few minutes if the meat sample contains sulfite. Allow ten minutes reaction time before deciding the test is positive.
5. The number of parts per million of sulfite equivalent to sulphur dioxide can be established by the following procedure.

    Prepare ground meat sample containing known levels sodium sulfite (Na_2SO_3)

    To 100 g of ground meat add 0.0146 g of 98 percent sodium sulfite (50.83 percent equivalent SO_2) and thoroughly mix the sample. The resulting sample mixture will contain equivalent 200 ppm of SO_2.

    Take 50 g of the 200ppm mixture and thoroughly mix with 50 g of ground meat, which has been not adulterated. The resulting sample mixture will contain equivalent 100 ppm of SO_2.

Prepare ground meat samples containing 50, 25, and 12 ppm of SO_2 according to same procedure. Prepare also a sample of meat containing no added sulfite to determine the sulfite that occurs naturally in fresh meat.

Conduct those 6 tests using lead acetate paper (0,12,25,50,100 and 200 ppm SO2) respectively. Also conduct a seventh test using no meat. Limit the time to each of the seven tests to 5 minutes. When dried, these test papers can be used as standards for comparison against your tested paper used with meats suspected to contain sulfite.

7.10 ESTIMATION OF CURED MEAT PIGMENT CONTENT

Principle

This procedure is based on the extraction of nitroso-pigments from cured meat products with an acetone - water solvent. With the inclusion of hydrochloric acid in the solvent the method can be adapted to measure the total pigments present.

Apparatus

1. Spectrophotometer - 1 cm cuvettes should be used.
2. Glass funnels 50-mm diameter.
3. Beakers, high form 100 ml capacity (no pour out)
4. Watch glasses to cover beakers
5. Glass stirring rods
6. Balance.

Reagents

1. Acetone, reagent grade,
2. Hydrochloric acid, specific gravity 1.19

Procedure

Test for Cured Pigment (Nitrosohematin)
1. Weigh 10 g of sample and place into high form beaker.
2. Mix and macerate the sample with solution consisting of 40 ml of acetone and 3 ml of distilled water.
3. Continue mixing the sample for another 5 minutes under reduced light.
4. Filter the solution through Whatman filter paper No. 1, or equivalent.

5. Measure optical density of prepared filtrate at 540 nm and multiply the absorption (optical density) by 290 to obtain nitroso hematin pigment in ppm (parts per million.) against blank containing 80percent acetone in water.

Test for Total Pigments

1. Weigh 10 g of sample and place into high form beaker.

2. Add 40 ml of acetone, 2 ml of water and 1 ml of concentrated hydrochloric acid.

3. Mix the sample well, cover with watch glass and keep for 1 hour in dark. Filter through Whatman filter paper No. 1.

4. Prepare spectrophotometer (with 1 cm cell) at 640 nm wavelength adjusted to zero with blank sample containing 80percent acetone,2percent HCl and 18percent distilled water.

5. Measure optical density of prepared filtrate with added HCl at 640 nm against blank sample containing 80 percent acetone, 2 percent HCl and 18 percent of water. Express total heme pigments in ppm by multiplying the optical density by 680.

Calculation

$$\frac{\text{ppm nitroso pigment} \times 100}{\text{ppm of total pigment}} = \text{Percent of conversion.}$$

References

American Meat institute, Laboratory Methods of the Meat Industry, Chicago, III. AOAC (1975)

Church, P.N. and Wood, J.M. (1992) In the manual of manufacturing meat quality, Elsevier Applied Science Publishers Limited, Crown house, Linton road, Barking, Essex, IG11 8JU, England.

Koniecko, E.S. in Hand book for meat chemists, Avery publishing group, Wayne, New Jessey (IS:5960 part VI, 1971.)

Kowale, B.N., Kulkarni, V.V. and Kesava Rao, V. (2008) Methods in meat science. *Pub.* by Jaypee brothers medical publishers (P) Ltd., Ansari Road, Daryagunj, New Delhi.

Laboratory Guidebook, Policy and Procedures of the laboratory Branch, U.S.D.A AOAC (1975).

Merck Products for the Meat Industry, Merck & Co., Incl., Rahway, New Jersey.

CHAPTER-8

■ ■ ■

TESTS FOR DETERMINATION OF
KEEPING QUALITY OF MEAT

8.1 MEASUREMENT OF EXTRACT RELEASE VOLUME (ERV)

The extract release volume is used for assessing the spoilage of beef. The procedure is based on measuring the volume of aqueous filtrate released from slurry of meat in a fixed time. The ERV decreases as the spoilage progresses and no filtrate at all is obtained with putrid meat.

Procedure

Take 25 g of minced meat and homogenise with 100 ml of 0.1 M phosphate buffer (pH 5.8) and the slurry is filtered through Whatman filter paper for 15 min. in a measuring cylinder.

The filtrate volume of 25 ml is taken as cut-off figure. The draw back of this estimation is that fairly wide range of values is given by fresh meats (21-35 ml).

In view of its simplicity, rapidity in performance and the apparently consistent decrease during spoilage, the ERV may well prove useful for routine control assessment of meat quality.

8.2 MEASUREMENT OF WATER ACTIVITY (a_w)

Water activity is the amount of water available for the growth of microorganisms in foods. a_w is defined as the ratio of the water vapour pressure of food substrate to the vapour pressure of pure water at the same temperature. The reduction in aw of food by drying, desiccation or by other means increases the keeping quality as the moisture level is below the level required for the growth of microorganisms. The a_w of most fresh foods is 0.99. The minimum a_w required for the growth of bacteria, molds and yeast are 0.80, 0.61 and 0.60, respectively. However most spoilage bacteria do not grow below a_w 0.91 while spoilage molds can grow at as low as 0.80 with respect to food poisoning bacteria. Staphylococcus aureus has been found to grow at as low as 0.86, while Clostridium botulinum does not grow below 0.94. Certain relationship exist between aw, temperature and nutrition. At any temperature the ability of microorganisms to grow is reduced as the a_w is lowered. The range of aw over which growth occurs is greatest at the optimum temperature and the presence of nutrients increases the range of aw over which the organisms can survive. The general effect of lowering aw below optimum is to increase the length of lag phase of growth and to decrease the growth rate and size of final population. This effect may be expected to result from adverse influence of lowered water on all metabolic activities since all chemical reactions of cells require an aqueous environment.

Salt –

Specific a_w of a number of common salts used are:

K_2SO_4	KNO_3	$BaCl_2$	KCl	$(NH_4)2SO_4$	$NaCl$
0.973	0.936	0.902	0.843	0.810	0.753

Measuring water activity using a_w meter: a_w is measured by simple user friendly operations involving minimal steps. Steps involved in measuring a_w using Water activity meter (Make: Decagon devices, Pullman, Washington, USA) is given below:

- Crush the samples and place it in sample holder provided with the equipment.
- Place the sample holder with sample inside the water activity meter taking proper care not to touch the sensor and closed upper lid and press the reading button.
- Once the reading is completed which is indicated by beep sound, take out the sample holder.
- It takes 4 to 5 minutes to complete one reading.
- Readings are usually taken in triplicates.
- Water activity meter need to be calibrated at regular intervals

8.3 MEASUREMENT OF ATP'ASE ACTIVITY

Introduction

ATP is the ultimate source of energy to muscle for the contractile process, for pumping of calcium during relaxation, and for maintaining the sodium and potassium gradients across the sarcolemma. This energy is derived from ATP in a reaction, catalysed by the enzyme myosin ATPase, in which the ATP is hydrolysed to adenosine diphosphate (ADP) and inorganic phosphate. ATPase activity is temperature dependent and even proceeds at a slower rate during low temperature storage of meat. The estimation of ATPase gives an idea of the extent of breakdown of ATP and progress of rigor mortis.

Equipment

Fine analytical balance, high speed blender (15000 rpm), Spectrophotometer.

Reagents

1. Meat homogenate – Homogenise one gram of meat sample with 10 ml of ice cold water in a vortex mixer and pass through cotton gauge to remove fascia.
2. Reaction mixture- consist of 1 M NaCl, 1.2 ml (120 mM), 1M KCl, 0.30 ml (30 mM), 1M $MgCl_2$, 0.05 ml (5 mM), bovine serum albumin 0.2 percent 0.1 ml and 3.35 ml glass distilled water. Use 0.25 ml per assay for a final volume of 0.5 ml, (pH 7.4 at 37°C).
3. 30 mM ATP- Dissolve 181.8 mg (disodium salt) in 6 ml water, adjust pH with 0.5 M tris base and make up to 10 ml with glass-distilled water. Keep frozen.
4. 50 percent TCA (W/W) in water

Procedure

Estimate protein content of meat homogenate by Lowry's method or Biuret method.

Take 0.25 ml reaction mixture in two test tubes.

Add 0.05 ml meat homogenate to both the tubes. Make the volume of incubation mixture to 0.45 ml with glass-distilled water.

Start the reaction at 37°C with the addition of 0.05 ml of ATP solution. Stop the reaction after 10 minutes by adding 0.05 ml of 50 TCA and immerse the tubes in ice bath for 10 minutes. Run enzyme blank in similar way but the enzyme is added after addition of TCA. Centrifuge the tubes for 5 minutes at 5000 rpm to remove precipitate and use whole supernatant for inorganic phosphorus (pi) assay by the method of Fiske and Subba Row (1925).

Calculation

The enzyme activity is expressed as m Moles of pi liberated / hr / mg protein. OD unknown / OD standard x concentration standard / mg of protein x 2 (Paul, 1975 and Fiske Subba Row, 1925)

8.4 MEASUREMENT OF R VALUE

Pulverise three g of liquid nitrogen frozen breast muscle sample in a blender for one minute. Now homogenise in 20 ml of 1 M perchloric acid using virtis macromodel homogeniser at 45,000 rpm for 1 min. and filter the homogenate through Whatman No. 4.

Add one tenth of an ml of the acid filtrate to 4 ml of 0.1 M phosphate buffer (pH 7.0). Use phosphate buffer (0.1 M. pH 7.8) as blank.

Read the absorbance at 250 and 260 nm for IMP and ATP.

R = Absorbance at 250nm / Absorbance at 260nm.

(Khan and Frey, 1971)

8.5 MEASUREMENT OF ACTIVITY OF CATHEPSIN B, D, H

An assay mixture for cathepsin D consists of 200ml of 10 percent tissue homogenate prepared in 0.1 M acetate buffer (pH 3.8), 300 ml of 0.1 M acetate buffer (pH 3.8) and 500ml 1percent haemoglobin prepared in the same buffer. The reaction is carried out at 50°C for 30 min and stopped by adding 1 ml 10 percent trichloroacetic acid (TCA). The TCA soluble products are determined using tyrosine as a standard. Activity of cathepsin B is determined using BANA (Na -benzoyl-DL-Arg b -napthylamide) at pH 6.2 at 40°C for 1 hour. Cathepsin H is assayed using BANA as substrate at pH 6.8 at 40°C for 30 min. Both the reactions were terminated by addition of 1ml Mersayl acid. The activity is determined by monitoring the release of b -napthylamine at 520 nm and expressed as m Moles / hr / g tissue) (Barrett, 1972; Jamadar and Hari Kumar, 2002).

8.6 ESTIMATION OF MUSCLE GLYCOGEN

Introduction

Glycogen (0.10percent) is an important muscle carbohydrate besides lactic acid (0.50 percent), glucose - 6 - phosphate (0.15 percent) and glucose, along with traces of other glycolytic intermediates (0.05 percent) out of a total of 2.5 percent. Glycogen plays a vital role in pre and post mortem glycolysis as a source of energy to muscle by its conversion to lactic acid due to action of various glycolytic enzymes. Amount of glycogen in muscle gives an indication of activity of muscles, extent and pattern of rigor mortis after the slaughter of animals. Glycogen content in the muscles is influenced by species, breed, type and activity of muscles, temperature and plane of nutrition.

Principle

Glycogen is hydrolysed to glucose and the glucose thus formed is estimated by any standard method.

Reagents

1. 30 percent KOH
2. 95 percent Ethanol, 60 percent Ethanol
3. 2 N H_2SO_4
4. Phenol (80 percent W/W)

   5. Standard glucose solution- High purity glucose is dried in a vacuum oven at 60-70°C.Dissolve 100 mg of glucose in 100 ml of saturated benzoic acid solution in a volumetric flask. For working standard use a solution in benzoic acid having 0.1 mg of glucose / ml.

Procedure

1. Weigh 5 g muscle taken out rapidly from the carcass. Remove excess blood by blotting between folds of filter paper and immediately put into a weighed stoppered test tube. Add 10 ml of 30percent KOH.
2. Digest the tissue in a boiling water bath for 30 minutes.
3. Cool in ice cold water. Add two volumes of 95percent ethanol and heat the mixture heated just to boiling.
4. Allow to stand overnight in the cold, centrifuge and dissolve the precipitate in 5-10 ml warm water. The glycogen is re-precipitated with 2 volumes of 95percent ethanol.
5. Centrifuge the precipitate at 1000 rpm for 10 minutes and wash several times with 60percent ethanol. Glycogen is directly estimated by estimating glucose content spectrophotometrically.
6. Dry the precipitate in vacuum desiccator. Dissolve it in distilled water and make the volume to 100ml.

7. Take 0.1 ml aliquot from the above solution in a test tube and add 1.9 ml distilled water to make 2 ml, add 0.1 ml phenol solution (80percent w/w) followed by 5 ml concentrated H_2SO_4. Read optical density at 490 nm. For blank take 2 ml distilled water. For standard use 5 to 30 mg in 2 ml volume.

Calculation

$$\text{Mg glucose } /100 \text{ g tissue} = \frac{\text{O.D.of unknown} \times \text{Conc. of standard} \times 20}{\text{O.D. of standard}}$$

The factor 0.93 is used to convert glucose value to glycogen content. (Montogomery Rex, 1957)

8.7 GLYCOGEN ESTIMATION BY ANTHRONE METHOD

Reagents

Anthrone reagent:- A solution containing 0.05 percent anthrone, 1 percent thiourea and 66 percent by volume H_2SO_4 is used. For each liter of reagent, place in a suitable flask 340 ml of distilled water and add cautiously 660 ml conc. H_2SO_4 (sp. gravity 1.84). Prepare a stock of this 66 percent H_2SO_4. Place in a flask 500 mg of crystalline anthrone. 10 g of highest purity thiourea and 1 liter of 66 percent H_2SO_4. Warm the mixture to 80 to 90°C, occasionally shaking the flask. Do not over heat. Cool and store in a refrigerator. To obtain maximum colour development, this reagent should be freshly prepared. Standard glucose solution: Obtain highest purity glucose and dry it in a vacuum oven at 60 to 70°C. Dissolve 100 mg of glucose in 100 ml of saturated benzoic acid solution. For working standard use a solution in benzoic acid having 0.1 mg of glucose / ml.

Equipment

Fine analytical balance, high speed blender (15000 rpm), Spectrophotometer.

Procedure

1. Weigh 5 g muscle taken out rapidly from the carcass. Remove excess blood by blotting between folds of filter paper and immediately put into a weighed stoppered test tube. Add 10 ml of 30percent KOH.
2. Digest the tissue in a boiling water bath for 30 minutes.
3. Cool in ice cold water. Add two volumes of 95percent ethanol and heat the mixture heated just to boiling.

4. Allow to stand overnight in the cold, centrifuge and dissolve the precipitate in 5-10 ml warm water. The glycogen is re-precipitated with 2 volumes of 95percent ethanol.

5. Centrifuge the precipitate at 1000 rpm for 10 minutes and wash several times with 60percent ethanol. Glycogen is directly estimated by estimating glucose content spectrophotometrically.

6. Dry the precipitate in vacuum desiccator. Dissolve it in distilled water and make the volume to 100ml.

7. Take 1 ml test solution. Add 1 ml distilled water and 1 ml standard glucose solution (0.1 mg glucose) in different colourimeteric tubes. Add 10 ml anthrone reagent and mix. Put the rubber stopper and place them in boiling water bath for 15 min. Allow it to cool and read at 620 nm. For standard use 5 to 30 mg in 2 ml volume.

Calculation

$$\text{mg glucose /100 g tissue} = \frac{\text{O.D. of unknown} \times \text{Conc. of standard} \times 2}{\text{O.D. of standard}}$$

The factor 0.93 is used to convert glucose value to glycogen content. (Rao, 1955)

8.8 ESTIMATION OF LACTIC ACID

Introduction

Lactic acid is produced in muscles during aerobic as well as anaerobic metabolism by breakdown of glycogen. Lactic acid accumulation in the muscles lowers its pH and at the pH values less than 6.0-6.5 the rate of glycolysis is drastically reduced. The accumulation of lactic acid early in the postmortem period can have an adverse effect on meat quality. Development of acidic condition in muscles, before the natural body heat and the heat of the continuing metabolism have been dissipated through carcass chilling, causes denaturation of muscle proteins. The amount of denaturation depends on how the high temperature and low pH have reached. Denaturation of muscle proteins causes a loss of protein solubility, loss of water holding capacity and a loss in the intensity of the muscles pigment colouration. Estimation of lactic acid, glycogen and pH can give a better picture of the development of rigor mortis in hot carcass and meat quality during the various periods of storage. For tissue analysis precautions must be taken against post mortem changes in lactic acid content by prompt freezing in solid CO_2 or by adequate treatment with acid to destroy enzymes present.

Principle

The glucose and other interfering material of protein free blood filtrate are removed by Van Slyke - Salkowski method of treatment with copper sulphate and calcium hydroxide. An aliquot of the resulting solution is heated with concentrated sulphuric acid to convert lactic acid to acetaldehyde, which is then determined by reaction with p- hydroxydiphenyl in the presence of copper ions.

For tissue analysis precaution must be taken against postmortem changes in lactic acid content, by promoting freezing in solid CO_2 or by adequate treatment with acid to destroy enzyme system present. Proteins may be removed by any of the common methods. An aliquot of protein free fluid containing 0.02 to 0.10 mg of lactic acid is treated by the copper lime procedure at a volume of 10 ml and 1 ml of supernatant is analysed as described above for blood.

Reagents

1. 20 percent copper sulphate solution: - Dissolve 400 g Copper sulphate in about 1 liter of water with the aid of heat, cool, dilute to 2 liters and mix. This solution is stable indefinitely.

2. 4percent copper sulphate solution: - Dilute 1 volume of 20 percent $CuSO_4$ solution to 5 volumes with water and mix. Store in a bottle fitted with a stopper carrying a 1 ml pipette, which delivers approximately 20 drops per ml. If this is done, one drop may be used instead of 0.05 ml portion specified in the text.

3. Calcium hydroxide powder

4. Concentrated H_2SO_4: - It is dispensed from burette, the stopcock cleaned thoroughly of grease and lubricated with a little of the acid itself.

5. P- hydroxydiphenyl reagent: - Dissolve 1.5 g of p-hydroxydiphenyl in 10 ml of 5 percent sodium hydroxide solution plus a little water, by warming and stirring and dilute to 100 ml with water. Store in a brown bottle fitted with a stopper and pipette capable of delivering 20 drops per ml. If this done 2 drops may be used instead of 0.1 ml portion. Reagent is stable for many months; deterioration is evidenced by high blank reading.

6. Standard lactic acid solution: -This is prepared preferably from lithium lactate, which is anhydrous. For the stock standard dissolve, 0.213 g of pure dry lithium lactate in about 100 ml of water in a one liter volumetric flask, add about 1 ml of concentrated H_2SO_4, dilute to the mark with water and mix. This solution contains 1 mg of lactic acid in 5 ml and is stable indefinitely

if kept in the refrigerator. To prepare working standard, dilute 5 ml of stock standard to 100 ml in a glass stoppered volumetric flask with water and mix. This solution contains 0.01 mg of lactic acid per ml, and is best prepared fresh.

Procedure

Weigh accurately two gram of muscle tissue and deproteinise with 10 ml of 20 percent TCA. Transfer 2 ml of the protein free filtrate, representing 0.2 g of muscle, to a centrifuge tube graduated at 10 ml. In a second similar tube place 5 ml of standard lactic acid solution, containing 0.01 mg of lactic acid per ml. In a third test tube place 1 ml water for a blank. To each tube add 1 ml of 20 percent copper sulphate solution and dilute to 10 ml mark with water. Add 1 g of powdered Ca $(OH)_2$ to each tube, put the glass stopper and shake vigorously until the solids are uniformly dispersed. Allow to stand for one and half hour, repeating the shaking at least once. Centrifuge down the precipitate and transfer in duplicate 1 ml of the supernatant from each tube to thoroughly dry and clean test tubes (diameter 18×23 mm). To each tube add 0.05 ml of 4 percent copper sulphate solution, followed by 6 ml concentrated H_2SO_4 from a burette, drop by drop and mixing the contents of the tube well during the addition .The tube contents will become hot; it is not necessary to cool the tube. After acid has been added to all the tubes, place them upright in boiling water for 5 minutes, then transfer the tubes to cold water [preferably running] and cool to 20°C or below. When the contents of the tubes are sufficiently cool, add 0.1ml of the p-hydroxydiphenyl reagent, drop by drop, to each tube .The reagent precipitates out on entering the concentrated acid; it is dispersed throughout the solution as quickly and uniformly as possible by lateral shaking. When the reagent has been added, place the tubes in a water bath at 30°C and allow to stand for 30 min. Finally place the tubes in vigorously boiling water for exactly 90 sec. Remove and cool to room temperature. Read optical density at 560 nm, using blank for setting the photometer at zero density.

Calculations

$$\text{mg lactic acid /100 g tissue} = \frac{\text{O.D. of unknown} \times \text{Conc. of standard} \times 100}{\text{O.D. of standard} \times 0.2}$$

8.9 ESTIMATION OF TYROSINE VALUE

Introduction

Tyrosine value is an indicator of proteolysis as it measures the amino acid tyrosine and tryptophan present in a non-protein extract of meat. Tyrosine value increases during storage but the increase is more evident with advanced spoilage than the changes occurring during early stage of spoilage. A significant correlation of tyrosine value with ammonia and NPN values and bacterial counts has been observed.

Equipment

Fine analytical balance, high-speed blender (15000 rpm), Spectrophotometer.

Reagents

1. 20 percent Trichloroacetic acid
2. Foiln Ciocalteau reagent (3 N): - Dissolve 100 g sodium tugnstate and 25 g of sodium molybdate in 700 ml of distilled water and add 50 ml of orthophosphoric acid (specific gravity 1.75) and 100 ml of conc. HCl. Reflux for 10 hours and to this add lithium sulphate (150g) distilled water 50 ml and few drops of Bromine water. Boil without condenser for further 15 min. cool and make up to 1 liter. Store in dark bottle. Dilute with two volumes of distilled water for making working solution.
3. 0.05 N NaOH

Procedure

Weigh 2 g of meat and add 10 ml of chilled 20percent TCA. Homogenise in a vortex mixer for 2 min. Allow to stand for 10 min. Filter through Whatmann filter paper No.42. Pipette 2.5 ml of filtrate and add 2.5 ml of distilled water and 10 ml freshly prepared 0.5 N NaOH. Mix and keep the tubes for 10 min. Add 3 ml of Foiln Ciocalteau reagent, wait for 30 min. and allow to develop colour in dark. Read the optical density at 730 nm.

Calculation

$$\text{Tyrosine (mg/100 g)} = \frac{\text{O.D. of unknown} \times \text{Conc. of standard} \times 10 \times 100}{\text{O.D. of standard} \times 2 \times 2.5 \times 1000}$$

(Strange *et al.*, 1977)

8.10 ESTIMATION OF TOTAL VOLATILE NITROGEN (MICRO DIFFUSION TECHNIQUE)

The volatile bases in most species of fish consist of ammonia together with appreciable quantities of amines. In meat trimethylamine is only present in insignificant quantities and total volatile nitrogen consist almost entirely of ammonia. As ammonia production due to de-amination of protein increases during spoilage, its determination represents a simple method of following the course of determination of the quality lean meat.

Principle

Meat extract is treated with relatively weak alkali and the volatile base is distilled or diffused over into standard acid or boric acid.

Reagent

Boric acid reagent - Dissolve 5 g Boric acid in 100 ml of 95percent alcohol and add 350 ml of water. After the acid had dissolved, add 5 ml of indicator (0.066percent methyl red and 0.33 percent bromocresol green in alcohol). Add alkali (40percent sodium hydroxide) until a faint reddish colour is produced. Make the volume up to 500 ml with alcohol.

Procedure

Homogenise 100 g minced meat sample with 2.5 to 5 g powdered trichloroacetic acid in a porcelain basin to make fine slurry. Cover with aluminum foil. Allow to stand for 30 min. Filter on a Buchner funnel through a Whatman paper no.5. Keep the filtrate under refrigeration until use. The diffusion is carried out in triplicate together with a blank of 2 ml of the Boric acid solution (added to the central compartment of a Conway micro-diffusion unit). Carefully pipette 1 ml of meat filtrate (2:10 dilution for spoiled sample) into the other compartment by sliding the lid (smeared with Dow -corning grease). Also add 1 ml of saturated potassium carbonate solution through the gap and slide the lid quickly to form an airtight seal. Rotate the dish mechanically on a Gallon kamp oscillating table and allow the diffusion of volatile nitrogen at 37°C for 3 hrs, during incubation rotate the dish 2-3 times. Titrate boric acid in the central compartment with 0.02 N H_2SO_4 and calculate the total volatile nitrogen as mg N / 100 ml of meat slurry. Adding known amount of ammonium chloride check recoveries.

Mg TVN / 100 g meat =

= Reading of burette × normality of acid used for titration × 14 × 100

= X × 0.02 × 14 × 100

(Pearson, 1968)

8.11 ESTIMATION OF NON-PROTEIN NITROGEN

The non-protein nitrogen (NPN) can be extracted from the water soluble fraction of muscle proteins as explained in the methods of Kang and Rice (1970).

Take 15 ml of sarcoplasmic fraction, mix with 5 ml of 10percent TCA and allow it to stand for 15 minutes. Filter through Whatman no. 1 paper. Analyse the nitrogen content in the filtrate by Kjeldahl method (Hegarty *et al.*, 1963; Kang and Rice, 1970).

8.12 ESTIMATION OF FREE AMINO ACID CONTENT

Free amino acid content is determined in the non-protein nitrogen (NPN) fraction by ninhydrin reagent.

Reagent

Ninhydrin reagent – Dissolve 0.3 g of ninhydrin in 97 ml of N-butanol and 3 ml of glacial acetic acid.

Procedure

To 10 ml of sarcoplasmic fraction of protein add 5 ml 10 percent TCA to precipitate the true proteins. Centrifuge the contents for 15 minutes and decant the supernatant. Dilute 2 ml of supernatant with distilled water adjusting the pH to 7 and making the final volume to 100 ml in a volumetric flask. From this take 2 ml and add 1 ml ninhydrin reagent and heat in a water bath (70°C) for about 10 minutes. Cool and measure the absorbance at 530 nm in a spectrophotometer. The concentration of free amino acids can be arrived by the regression equation constructed from reference value.

Reference value: - Dissolve 65.5 mg of Leucine in 100 ml distilled water. From this transfer 0.5, 1, 1.5, 2.0 and 2.5 ml into 100 volumetric flask and follow the steps as given above. Plot a graph for the known concentration of Leucine and construct a regression equation (Rosen, 1957).

8.13 MEASUREMENT OF OXIDATIVE RANCIDITY IN FAT

Rancidity is perceived as an unacceptable taste, and although many chemical and physical tests have been developed for the assessment of oils, fats and fatty foods, none of these measures rancidity itself. There are two types of rancidities: hydrolytic and oxidative. In case of hydrolytic rancidity the off flavour is caused by free fatty acids, the concentration of which can be easily measured. However, it is the composition of these acids as well as their level that determines the extent of off flavour perceived on the palate. Peroxide value measurement is the most widely accepted parameter for the measurement of oxidative rancidity in fats but peroxide themselves are tasteless and odourless. The determination of peroxide value of a fat is only an indirect guide to a product flavour. Several tests measure levels of the secondary oxidation products, that is the aldehyde and ketones produced when peroxide initially formed decomposes. The best known of these is the anisidine value. It must also be remembered that off flavours and off odours may be dues to causes other than rancidity.

If rancid fat are incorporated into product it causes browning and the product will have objectionable flavours and odours. To minimise the risk of receiving rancid fats, maximum storage time for fats should be specified and the fats prone to rancidity must be vacuum packed. It is important to know the time for which fat has been stored because peroxide values increase, as the fat becomes rancid but then decline with the extended storage. Therefore, fat with low peroxide value could be rancid.

Anisidine value (AV) is used to measure oxidative rancidity. P- anisidine gives yellow colour with carbonyl compounds, which can be measured and probably gives better correlation than TBA value with the breakdown of fat. The AV values below 5 are often specified. The anisidine value (AV) is often used in conjunction with the peroxide value to calculate the so-called total oxidation value of oil, referred as Totox value.

Totox value = 2 PV = AV (value below 10 are often specified).

8.14 ESTIMATION OF FREE FATTY ACID CONTENT

8.14.1 Estimation of Free Fatty Acid Content by Titration Method

Fatty acids present in substantial amounts in fats of meat animals are oleic, palmitic and stearic. Hydrolytic rancidity in fats is generally caused by a combination of microorganisms and moisture. Free fatty acids are formed by the action of enzymes-lipases, microorganism or by the catalytic action of trace metals such as copper, iron and nickel, hence goes on increasing during storage of meat. Free fatty acids, which are the products of enzymatic or microbial lipolysis of lipids, are responsible for flavour determination in meat, which is particularly noticeable when the acidity of the extracted fat reaches 2 - 3 percent (as oleic acid). The FFA values below 0.5 percent oleic acid are often specified. Fats with FFA values greater than 3percent are considered inedible.

Principle

A sample of fat is mixed with neutralised ethanol and titrated directly with standard alkali using phenolphthalein as indicator. The FFA is expressed as percent oleic acid.

Equipment and reagent

Titration assembly, fine analytical balance, 0.1 N NaOH.

Procedure

Pipette a suitable aliquot of lipid extract in 5 ml of freshly neutralised alcohol and titrate with 0.1 N NaOH using phenolphthalein as indicator. Use palmitic acid as standard.

Calculation

Express FFA content as g / 100 g tissue

$$\text{FFA as palmitic percent} = \frac{\text{ml of alkali} \times \text{normality of alkali} \times 25.6}{\text{Weight of the sample}}$$

8.14.2 ESTIMATION OF FREE FATTY ACIDS IN MEAT BY TITRIMETRIC AOAC METHOD

Extraction

One of the important steps in this procedure use of slow speed blending to get a clear extract. Slow blending extracts the fat but not water especially in low fat high water meat samples.

Weigh 50 g meat on a watch glass and transfer in the blender. Add half teaspoonful of anhydrous Sodium sulphate (Na_2SO_4) and about 137 ml chloroform

(CHCl$_3$). Blend for about 2 min and filter through 12 No. Whatman fluted filter paper. Collect the filtrate in 250 ml stoppered flask. Repeat the extraction by adding 50 ml of chloroform (CHCl$_3$) and transfer to the flask. Adjust the volume to 250ml.

Procedure

Pipette 25 ml filtrate in a 125 ml flask and add 10 drops of phenolphthalein indicator. Titrate with 0.01 N alcoholic Potassium hydroxide (if the sample is expected to have low FFA content) or with 0.1 N alcoholic Potassium hydroxide (if the sample is expected to have high FFA content) to end point.

Calculation

$$\text{Percent FFA as oleic acid} = \frac{\text{Normality of alkali} \times \text{ml of alkali} \times 28.2 \times 10 \times 100}{\text{Meat weight}}$$

8.15 ESTIMATION OF CARBONYL COMPOUNDS

Reagents

2-4 dinitrophenylhydrazine (DNPH) Reagent- Dissolve 2 g DNPH in 1 lit of 2N HCl

Procedure

Homogenise 10 g meat with 40 ml of water in a waring blender for 15 sec at low speed. Transfer the homogenate in a glass stoppered flask rinsing with 50 ml of DNPH. Shake well and allow to stand for 2 hrs permitting the formation of total carbonyl DNP derivatives (2-4 dinitrophenylhydrazones). Extract these derivatives with 60 and 40 ml of purified N Hexane. Read optical density of hexane extract at 340 nm in a spectrophotometer.

Calculate total carbonyl DNP derivatives from the absorbance using a molar extinction coefficient of 22500 (Lawrence, 1965).

8.16 ESTIMATION OF THIOBARBITURIC ACID VALUE (TBA VALUE)

Oxidative rancidity and discolouration are very serious problems during storage and marketing of meat and meat products. The TBA value is extensively used as an empirical measure of the deterioration of fatty foods. The TBA test measures the aldehyde residues resulting from lipid peroxidation and deterioration in extractable and non-extractable lipids. The reported threshold levels of TBA is 1-2 mg of malonaldehyde / kg for rancidity in meat. A significant negative correlation of TBA value for bacterial count has been observed in the beef and buffalo meat during refrigerated storage and hence, it is concluded that TBA test may not be used for monitoring deteriorating changes produced by bacteria. Sudden decrease of TBA value during storage was attributed to the possible reaction of malonaldehyde with proteins, amino acids, amino containing phospholipids and malonaldehyde itself to form fluorescent compound resulting in reduced extraction of malonaldehyde.

8.16.1 Estimation of TBA Number by Filter Paper Method

Principle

2-Thiobarbituric acid combined with glacial acetic acid is used to develop a colour in extracts of meats and meat products that will establish TBA number, defined as malonaldehyde, which determines oxidative rancidity level.

Equipment

Fine analytical balance, high speed blender (15000 rpm), Spectrophotometer.

Reagents

1. Malonaldehyde tetraethoxy propane (TEP) standard: Dissolve 0.3055 g of TEP (1,1,3,3, tetraethoxy propane) in 100 ml 95 percent alcohol. This stock solution will contain 1 mg Malonaldehyde / ml. For working standard we use this as 0.2 to 1.0 mg / ml concentration.
2. 0.1 percent TBA reagent in water

Standard curve

Use working standard in the concentration ranging from 0.2 to 10 mg Malonaldehyde / ml

Procedure

Weigh 2 g of meat. To this add 10 ml of chilled 20 percent TCA. Homogenize in a vortex mixer for 2 min. Allow to stand for 10 min. Filter through Whatman

Filter Paper No.42, pipette 3 ml of filtrate and add 3 ml of 0.1 percent TBA reagent. Mix and place the tubes in boiling water bath for 35 min. Cool and read optical density at 530 nm.

Calculation

$$\text{Malonaldehyde (mg/kg)} = \frac{\text{O.D. of unknown} \times \text{Conc. of standard}}{\text{O.D. of standard}} \times \frac{10 \times 1000}{3 \times 2}$$

(Witte *et al.*, 1970)

8.16.2 Estimation of TBA Number by Distillation Method

Principle

Oxidation of fat gives products such as aldehydes, which are be separated by steam distillation and form red chromogen with 2 thiobarbituric acid.

Apparatus

1. Kjeldahl distillation rack
2. Waring Blender
3. Hot plate
4. Hot and cold water baths
5. Pipettes, measuring cylinders: 50, 100 ml
6. Volumetric flasks: 250, 500 ml
7. Spectrophotometer - 1 cm cuvette diameter.
8. Glass stopper flasks: 50 ml volume (Erlenmeyer).

Reagents

1. 2-Thiobarbituric acid, M.W. 144.15
2. Glacial acetic acid 99.5percent
3. Hydrochloric acid, ACS grade (36.5, 38.0percent)
4. Sulfanilamide, (A.R.)
5. 1,1,3,3 - Tetra-Ethoxy Propane (TEP) Reagent (only in case if standard solution is desired)

Solutions

1. TBA Reagent: Weigh 1.442 g of thiobarbituric acid and dissolve in 450 ml of glacial acetic acid, and make the volume with H_2O to 500 ml. Keep stored in dark bottle in refrigerator.
2. Hydrochloric acid solution: Mix 1 volume of acid with 2 volumes of distilled water.
3. Sulfanilamide reagent: Dissolve 1 g of sulfanilamide in solution containing 40 ml of concentrated HCl and 160 ml of distilled water.
4. TEP standard: Weigh 0.3055 g 1,1,3,3, Tetra-Ethoxy Propane in 100 ml of 95 percent alcohol. This solution can be kept refrigerated for 1 week. This solution contain 1 mg malonaldehyde / ml.
5. Working standard: - It ranged from 0.2 - 1 mg malonaldehyde / ml.
6. Anti-foam beads
7. Beads, stones or zinc granules (Mesh 20)

Procedure

A. Uncured Meats

1. Blend 10 g of prepared meat sample with 50 ml of distilled water in blender for 2 minutes. Transfer quantitatively to the Kjeldahl flask, using 47.5 ml of additional water for washing. Add 2.5 ml HCl (1:2) solution.
2. Add a small amount of anti-foam agent to Kjeldahl flask and several beads or some zinc to prevent bumping.
3. Heat the flask and collect 50 ml of distillate into graduated cylinder instead of receiving flask. (Time required should be about 10 minutes.)
4. Mix distillate well and pipette 5 ml into glass stoppered flask and add 5 ml of TBA reagent.
5. Mix and immerse in boiling water bath for exactly 35 minutes. Prepare blank consisting of 5ml of 10 percent TCA and 5 ml of TBA and place together with other samples in water bath for 35 minutes.
6. Cool in tap water for 10 minutes and read in spectrophotometer at 532 nm wavelength with instrument calibrated to read 0 percent OD with blank TBA-water sample. Multiply optical density by 7.8 to convert to mg of malonaldehyde per 1000 g of meat.

B. Cured Meats

1. Blend 10 g of sample with 49 ml of distilled water and 1 ml of sulfanilamide reagent. Use 48 ml of distilled water for washing the blender and add 2.0 ml of HCl solution.
2. Proceed from step 2 as indicated in the procedure for uncured meats.

Standard Curve

Prepare a standard curve from TEP solution that contains 1 mg malonaldehyde / ml. TEP per 1 ml solution stored at about 4°C and warmed up to room temperature just prior to use.

1. Pipette 0.2 to 1.0 μ g malonaldehyde / ml TEP solution into test tube.
2. Add TCA bring to 5 ml volume.
3. Prepare blank sample using 5 ml of 10 percent TCA only.
4. Add 5 ml of TBA solution - reagent to all, including blank.
5. Warm up in hot water bath for 35 min.
6. Cool in tap water and proceed to determine optical density at 532 nm.
7. Multiply optical density by 7.8 to convert to mg of malonaldehyde per 1000 g of meat.

    TBA value expressed as mg malonaldehyde / 1000 g of tissue.
    (Tarladgis *et al.*, 1960)

8.17 ESTIMATION OF TOTAL ACIDITY

Principle

Total acidity means the amount of standardised sodium hydroxide used to neutralise the acidity of submitted meat sample to the pH level of phenolphthalein used as indicator.

Apparatus

1. Balance
2. Volumetric flask
3. Waring Blender
4. Filter paper: Whatman No.42
5. Pipette, measuring with 0.1 ml divisions

Reagents and Solutions

1. Sodium hydroxide solution in water, 0.1 N accurately standardized.
2. Phenolphthalein indicator (1percent in alcoholic solution).

Procedure

1. Weigh 10 g of meat sample.

2. Blend with 200 ml of distilled water and make up in the volumetric flask to 250 ml mark with H_2O and filter through Whatman paper No. 42.

3. Collect 25 ml of filtrate that will represent 1 g of sample analysed.

4. Add 75 ml of distilled water and 3 drops of phenolphthalein indicator.

5. Titrate with 0.1 N of sodium hydroxide solution.

6. Sometimes distilled water shows pH 5.5 or so. You may prepare a blank sample by pouring 100 ml of distilled water which is exactly how much was used for sample analysed, put 3 drops of same indicator and titrate blank sample with 0.1 N NaOH to the end point of used indicator. Deduct from total reading for accurate results.

7. Express the number of used ml of 0.1 N NaOH to neutralize 1 g of sample.

8.18 MEASUREMENT OF HYDROPEROXIDE VALUE

Lipid hydroperoxides were extracted from 1 g surimi gels with 10ml chloroform /methanol (2:1).After centrifugation (5min at 2000g), an aliquot (2ml) of lower chloroform layer was mixed with an additional 1.3ml chloroform /methanol (2:1) and then reacted with 16.7µl each of 3.94M ammonium thiocyanate 0.072M ferrous chloride. Absorbance was measured at 500 nm after 20min of incubation, and LOOH were quantified on the basis of a standard curve prepared from cumene-hydroperoxide.

8.19 MEASUREMENT OF TOTAL PHENOLICS USING FOLIN-CIOCALTENS METHOD

The antioxidant extract or meat and meat products may be analyzed for total phenolics using the Folin–Ciocalteus (F–C) assay (Escarpa & Gonzalez, 2001) with slight modifications as per Naveena *et al.* (2008). Five gram of raw/cooked meat was homogenized with 25 ml of 70% acetone and kept overnight for extraction at refrigeration temperature. For antioxidant extract or pure solution, sample may be taken directly. Suitable aliquots of extracts were taken in a test tube and the volume was made to 0.5 ml with distilled water followed by the addition of 0.25 ml F–C (1 N) reagent and 1.25 ml sodium carbonate solution (20%). The tubes must be vortexed and record the absorbance at 725 nm after 40 min incubation. The amount of total phenolics was calculated as tannic acid equivalent from the calibration curve using standard tannic acid solution (0.1 mg/ml).

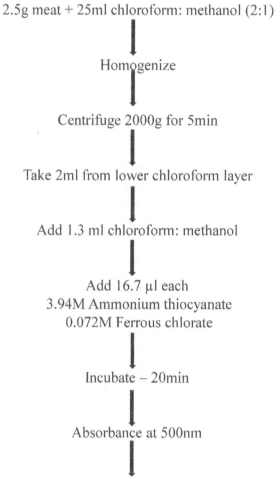

2.5g meat + 25ml chloroform: methanol (2:1)

Homogenize

Centrifuge 2000g for 5min

Take 2ml from lower chloroform layer

Add 1.3 ml chloroform: methanol

Add 16.7 µl each
3.94M Ammonium thiocyanate
0.072M Ferrous chlorate

Incubate – 20min

Absorbance at 500nm

Absorbance multiplied by 0.148 result in mg equivalent/kg sample.

Reagents

1. **Folin – ciocaltens reagent (1N):**

   Dilute commercially available Folin- ciocalten reagent (2N) with an equal amount of distilled water. Transfer it in a brown bottle and store in a refrigerator (4°C).It should be golden in colour. Do not use if it turns olive green.

2. **Sodium carbonate: 20%**

   Weigh 40gr of sodium carbonate ($10H_2O$), dissolve it in about 150ml distilled water and make up to 200ml with distilled water.

3. Std. tannic acid solution (0.1mg/ml)

    Dissolve 25mg Tannic acid obtained from merk in 25ml distilled water and then dilute to 1:10 in distilled water (Always use freshly prepared solution).

Antioxidants extracts (0.1, 0.25, 0.5ml)
diluted samples

↓

Makeup volume upto 0.5ml with distilled water

↓

Add 0.25ml FC reagent +1.25ml sodium carbonate solution

↓

Vortex & keep for 40min in dark

↓

Record absorbance at 725nm

Preparation of calibration curve

Tube Number	Tannic Acid Solution (ml) (0.1mg/ml)	Distilled water	Folin Reagent	Sodium carbonate Solution	Tannnic acid (μg)
Blank	0.00	0.50	0.25	1.25	0.00
1	0.02	0.48	0.25	1.25	2
2	0.04	0.46	0.25	1.25	4
3	0.06	0.44	0.25	1.25	6
4	0.08	0.42	0.25	1.25	8
5	0.10	0.40	0.25	1.25	10

Analysis

Take suitable aliquots of Tannin containing extract (initially by 0.02,0.05 and 0.1ml extract) in test tubes make up the volume to 0.5ml with distilled water and add 0.25ml F-C reagent and then 1.25ml sodium carbonate solution, vortex the tube and record absorbance at 725nm after 40min. Calculate the amount of total phenolics as tannic acid equivalent from the calibration curve.

Total phenolics (Tannic acid) standard curve

Tannic Acid (µg)	Abs 725nm
2	0.117
4	0.222
6	0.345
8	0.454
10	0.557

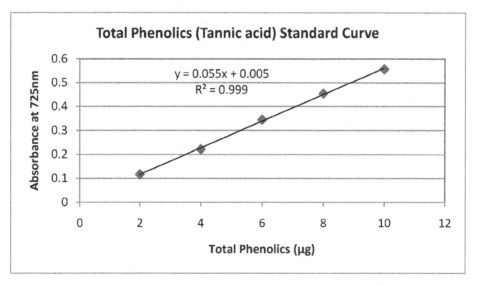

8.19.1 DETERMINATION OF REDUCING POWER

The reducing power of the test samples (antioxidant extract) was determined by the method of Jayaprakasha *et al.* (2001). Different concentrations of test samples (100 or 200mg) in 1ml MeoH were mixed with 2.5 ml of PBS (0.2M, P^H 6.6) and 2.5 ml of 1% pot. ferricyanide in a 10ml test tubes. The mixtures were incubated for 20 min at 50°C. At the end of incubation, 2.5ml of 10% TCA was added to mixtures and centrifuged at 5000 rpm at 10min. The upper layer (2.5ml) was mixed with 2.5ml distilled water and 0.5 ml of 0.1% ferric chloride, and the absorbance was measured at 700nm. The reducing power tests were run in triplicate. Increase in absorbance of the reaction indicated the reducing power of the samples.

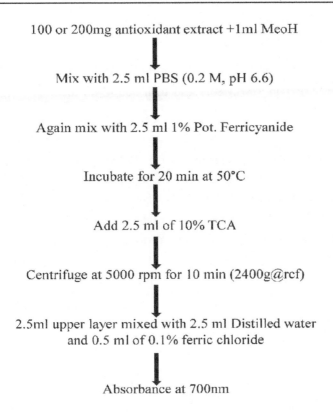

100 or 200mg antioxidant extract +1ml MeoH

↓

Mix with 2.5 ml PBS (0.2 M, pH 6.6)

↓

Again mix with 2.5 ml 1% Pot. Ferricyanide

↓

Incubate for 20 min at 50°C

↓

Add 2.5 ml of 10% TCA

↓

Centrifuge at 5000 rpm for 10 min (2400g@rcf)

↓

2.5ml upper layer mixed with 2.5 ml Distilled water
and 0.5 ml of 0.1% ferric chloride

↓

Absorbance at 700nm

8.19.2 DPPH Radical scavenging activity

The ability to scavenge DPPH radical by antioxidant extract was estimated by the method of Yamaguchi et al. (1999). Antioxidant extracts (0.5, 1.0, 1.5, 2.0 mg) in 1ml of 0.1 M Tris Hcl buffer (pH 7.4) was mixed with 1ml DPPH (250 μM) (1,1 –Diphenyl – 2 – Picrylhydrazyl) with vigorous shaking. The reaction mixture was stored in the dark at room temperature for 20min and the absorbance measured at 517nm.The scavenging activity was calculated by the following equations:

$$\text{Scavenging activity } \% = \frac{(\text{Absorbance blank} - \text{Absorbance samle})}{\text{Absorbance blank}} \times 100$$

Preparation of 250 μM DPPH

DPPH: MW 394.3

1M - 394.3 g in 1 litre

0.1 M (100mM) – 39.43 g in 1 litre

0.01 M (10 mM) - 3.943 g in 1 litre

0.001 M (1 mM or 1000 μM) – 0.3943 g in 1 litre

250 μM = 0.098575 g in 1 litre

For 50ml - 0.00493 g or 4.93 mg in 50ml Distilled water.

First dissolve DPPH in 1ml MeoH and then add 49ml distilled water.

References

Barker, S.B. and W.H. Summerson (1941) the colourimetric determination of lactic acid in biological material, J.Biol.Chem 138: 535. Modified by Huckabec, J. Appl. Physiol, (1956) 9,163.

Barrett, A.J. (1972) In Lysosomes – a laboratory handbook, Dingle J.T. (editor), North Holland Publishing Co. New York, USA.

Bureau of Indian Standards (BSI) have also suggested the method for plate count of bacteria in food stuff as per I.S. 5402: 1969.

Escarpa, A., & Gonzalez, M. C. (2001). Approach to the content of total extractable phenolic compounds from different food samples by comparison of chromatographic and spectrophotometric methods. Anal Chim Acta, 427, 119–127.

Fiske, C.H. and Subba Row (1925) J.Biol.Chem., 66:375-400.

Hegarty, G.R.; Brestzler, L.J. and Pearso, A.M. (1963) J. Food Sci. 28:525.

Jay, J.M. (1964) Fd. Technol. 18: 1633.

Jayaprakasha G.K., Singh, R.P., Sankaraiah, K.K. (2001).Antioxidant activity of grape seed (vitis vinefera) extracts on peroxidation models in vitro. Food chemistry, 73: 285-90.

Kang, C.K. and Rice, E.E. (1970) Journal of Food Science, 35: 563-565.

Khan A.W. and Frey, A.R. (1971), A simple method for following rigor mortis development in beef and poultry meat Can. Inst. Fd. Technol., J. 4: 139.

Koniecko, E.K. (1979) In Hand book for Meat Chemists, pp 68-69, Avery Publishing Group Inc. Wayne, New Jersey, USA

Lawrence, R.C. (1965) Nature 205: 1313 and Schwartz, D.P. Haller, H.S. and Kenney, M (1963) Anal. Chem, 35, 2191.

Modified AOAC, (1975)

Montogomery Rex (1957), Arch. Biochem. Biophys., 67: 378.

Naveena, B.M., Sen, A.R., Vaithiyanathan, S., Babji, Y. and Kondaiah, N. (2008). Comparative efficacy of pomegranate juice, pomegranate rind powder and BHT in cooked chicken patties. *Meat Science*, 80(4), 1304-1308.

Negi, P.S. and Jayaprakasha, G.K. (2003). Antioxidant and antibacterial activities of punicagranatum peel extracts. Journal of Food science, 68(4):1473-1477.

P.N.Church and J.M. Wood (1992) In the manual of manufacturing meat quality, Elsevier Applied Science Publishers Limited, Crown house, Linton road, Barking, Essex, IG11 8JU, England.

Park Y., Kelleher,S.D., Mc clement, J. and Decker, E.A. (2004).Incorporation and solubalization of omega-3 faty acids in surimi made from cod, gadusmorhua. J. Agric. Food chem. 52:597-601.

Paul Ottolenghue (1975) Biochem. J., 151, 61-6.

Pearson D (1968), Application of chemical methods for the assessment of beef quality. J. Sci. Food Agric., 19_pp 637.

Roe, J.H. (1955) The determination of sugar in blood and spinal fluid with anthrone reagent, J.Biol Chem. 212: 335- 343.

Rosen, H (1957) Arch.Biochem-Biophysics, 67: 10-15.

S.N.Jamadar and Harikumar, P (2002), J.Fd.Sci.Technol., 39(1) 72-73.

Strange, E.D., Benedict, R.C., Smith, J.L. and Swift, C.E. (1977) J. Fd. Protect., 40(12): 843-847.

Tarladgis, B.G., Watts, B.M., Younathan, M.T. and Dugan, L.R. (1960) J.Am. Oil. Chem. Soc.37: 44 - 48.

Witte, V.G., Krause, G.F. and Baily, M.E. (1970), J. Food Sci. 35: 582- 585.

Yamaguchi, T., Takamura,H., Matoba, T. and Terao, J. (1998). HPLC method for the evaluation of the free radical- scavenging activity of foods by using 1, 1 – Diphenyl – 2- picrylhydrazyl. Biosci. Biotechnol. Bochem, 62(6): 1201-4.

■ ■ ■

CHAPTER - 9

QUICK TESTS USED FOR DETECTING MEAT QUALITY

9.1 MALACHITE GREEN TEST FOR EVALUATING EFFICIENCY OF BLEEDING

The test is based on the principle that oxidised haemoglobin on addition of acidic solution of malachite green produces a green coloured complex. This test can be used to know the bleeding efficiency and also to identify meat from cold slaughtered (dead) animal, from that of meat obtained healthy live animal. In the case of cold slaughtered meat, there is high blood content in the subcutaneous blood vessels (due to minimum or no bleeding), the amount of oxidised haemoglobin content is more which forms a green coloured complex with acid malachite green.

Solutions

1 percent aqueous malachite green solution. This is diluted to 0.1percent with distilled water and pH is adjusted to 5.3.

Meat extract can be prepared by macerating one part of meat with two parts of distilled water and supernatant is used for test. Even the test can be performed directly on 1 to 2 g of meat sample.

Procedure

To the meat sample or meat extract 0.4 ml malachite green solution is added. Development of green coloured complex indicates inefficient bleeding. More darker the greenish colour less is the bleeding efficiency. The colour developed is stable for 18 hours; however, addition of hydrogen peroxide reduces the stability of colour (Shylaja *et al.*, 2001).

Alternate procedure

1. Preparation of malachite green reagent

    100 mg of malachite green is mixed with 25 ml of 30 percent solution of glacial acetic acid. The dye is dissolved by heating the mixture on water bath at 60°C for 10 min. After dissolution, 100 ml distilled water is added. It is stored in stoppered coloured bottle.

2. Hydrogen peroxide solution (3 percent)

3. Meat extract:- 6 g piece of muscle tissue removed with a sharp knife is placed into a small Erlenmeyer flask to which 14 ml of distilled water is added and is left for 15 minutes. The flask is shaken well. The mixture is then poured into a test tube for meat to sediment and to obtain a clear supernatant fluid.

Procedure

For the test, 0.7 ml of clear meat extract is taken in an agglutination tube and one drop of acidic malachite green solution is mixed. After mixing, 1 drop of hydrogen peroxide is added and the tube is shaken till foam develops. The tube is allowed to stand for 20 minutes for the development of colour.

A clear blue colour indicates normal bleeding; cloudy and green colour reaction shows imperfect bleeding, while a cloudy fluid of olive green colour shows very unsatisfactory bleeding (emergency slaughter of dead animal).

9.2 JAUNDICE TEST (RIMINGTON AND FOWRIE TEST)

Two g of fat to be tested is taken in a test tube and 5 ml of 5 percent NaOH is added to it and boiled for one minute, shake frequently and then cool under the tap. To it, is added equal volume of ether and allowed to settle.

If bilirubin is present, a water-soluble sodium salt is formed which is deposited at the bottom and is of greenish colour. If physiological yellow colour of fat is present the upper layer is yellow, while with both bilirubin and other pigment (carotene) an upper yellow and bottom yellowish green layers are formed.

9.3 WATER PROTEIN RATIO

Water protein ratio in muscle is generally constant and not affected by the fat content. In extremely emaciated animals, the percentage of water present in the muscle is about 80 percent and protein 19 percent giving a water protein ratio of 4:1, in lean meat water is 76.5 percent and protein about 22 percent, and the water protein ratio is less than 4:1. This test is very useful to differentiate between lean and emaciated animals.

9.4 BOILING TEST FOR DETECTING ABNORMAL ODOUR

A pan is filled with cold water and a piece of meat about the size of palm of the hand is put in it. Lid is replaced and is put on the flame for boiling.

When boiling starts, the lid is removed and odour is smelled (odour is most pronounced when water starts boiling). If abnormal odour is present the meat also has abnormal taste and if necessary the meat and water should be tested.

9.5 FRYING TEST

This test is suitable for determining the suitability of fatty tissues for marketing. When testing lean meats, a small amount of fat may be added previously to the frying pan. Rancid fat will give off odour on frying.

9.6 RESAZURIN TEST (DYE REDUCTION TEST)

This test estimates bacterial population in meat sample indirectly.

Four resazurin tablets are dissolved in 100 ml of water. Filter paper strips are dipped in the above solution and dried in a dark and cool room. For testing, the strip is moistened and a drop of meat juice to be tested is placed on it for 1 minute. The strip is then removed and placed in polythene bag and is kept in a dark room (22-23°C).

Time taken for the blue colour of the paper to change to red is noted and the results are interpreted as follows:

10 minutes: - meat is not acceptable

10-30 minutes: - may be acceptable

30-60 minutes: - good quality

over 60 minutes: - very good quality

9.7 NITRAZINE –YELLOW TEST

This test determines the acidity of meat.

Take a piece of meat free of blood, fat and connective tissue in a petridish and add nitrazine yellow indicator (1:10000) sufficient to cover the meat. Mix with a stirring rod. Note the colour change with the standard chart provided.

pH	Colour	Inference
6.0	Yellow	Good keeping quality
6.4	Olive green	Not having same good keeping quality
6.8	Bluish violet	Suspect or signs of incipient spoilage

9.8 COPPER SULPHATE TEST

This test is done to detect the staleness of meat

Take 10 g of minced meat and add 30 ml of distilled water, boil and filter through cotton. Cool to room temperature and take 5 ml of the filtrate and add 4 drops of 1 percent $CuSO_4$ solution.

Clear broth indicates fresh meat.

Turbid broth indicates doubtful.

Deep turbid broth indicates stale or unfit for consumption.

9.9 SODIUM NITRO-PRUSSIDE TEST

This test is done to determine metabolic disturbances such as acetonaemia / ketosis in animals or milk fever.

Take 10 g of minced meat each in two test tubes. Add 15 ml. distilled water in each tube. Add about 3 g of test reagents (a mixture 100g ammonium sulphate, 50 g of sodium carbonate and 3 g of sodium nitroprusside). Shake well and allow to stand for 3 minutes.

Interpretation

Purple colour indicates presence of ketones. Whenever the glucose level in blood plasma declines (e.g. starvation) concentration of free fatty acids in the plasma rises. This rise is paralleled by increase in the concentration of ketone bodies, which provide a normal physiological third source of energy. The carcass possesses a sweet odour, so the liver should always be incised and smelled in suspect cases.

9.10 MEAT SWELLING CAPACITY

This test determines the freshness of meat. Swelling capacity of meat increases during spoilage due to protein degradation and penetration of more amounts of water in protein matrix.

Blend 25 g of meat with 100 ml of distilled water for two minutes. Centrifuge 35 ml of homogenate at 2000 rpm for 15 minutes. Measure the amount of supernatant and pellet.

9.11 ODEMA TEST

Odema or dropsy denotes the presence of abnormal amounts of fluid in the tissues or the body cavities. In anacerca involving the subcutaneous and connective tissues, or in any form of odema accompanied by emaciation, the carcass should be condemned. In healthy cattle the bone marrow contains not more than 25 percent water, while if anasarca is present it contains more than 50 percent.

Procedure

Take a pea size piece of long bone marrow and put it in a beaker containing 32 percent, 47 percent and 52 percent alcohol.

Marrow containing 25 percent water will float in each case. If the water content of bone marrow is more than 50 percent, the carcass is unfit for food. Judgement of water content between 25 and 50 percent depends on the regulations of the particular country, such as for conditional release of the carcass.

Reference

Shylaja, R., Shobha,D., Bhagirathi, B., Radhakrishnan K, and Arya, S.S,. (2001). J. Food Sci, Technol.38 (4) 390-392.

PHYSICO-CHEMICAL PROPERTIES OF OILS AND FATS

10.1 MEASUREMENT OF SPECIFIC GRAVITY OF FATS AND OILS

The oil or fat sample may be preserved under an atmosphere of nitrogen in the cold. The sample should be allowed to reach the room temperature and mixed well before being subjected to the various analytical tests.

The specific gravity of a liquid is the weight of a given volume of liquid at the specified temperature, compared with the weight of an equal volume of water at the same temperature.

The specific gravity bottle (or pyrometer) is weighed empty (WB).

The bottle is then filled completely with the liquid and weighed (WL).

After cleaning, the bottle is filled completely with distilled water and weighed (WW).

The temperature of the liquid is noted.

$$\text{Specific gravity} = \frac{W_L - W_B}{W_W - W_B}$$

The temperature at which the specific gravity is determined should be specified (British Pharmacopoeia Appendix IV - E, p. 1017 (1963)).

10.2 MEASUREMENT OF REFRACTIVE INDEX

The refractive index (n) of a substance is the ratio of the velocity of light in vacuum to its velocity in the substance. It varies with the wavelength of light used in its measurements. It may also be defined as the ratio of the sine of angle of incidence to the sine of the angle of refraction.

Clean the refractometer with alcohol and ether.

A drop of oil or fat (in case of a solid fat the temperature should be suitably adjusted by circulating hot water) is placed on the prism. The prism is closed by the ground glass-half of the instrument. The dispersion screw is adjusted so that no colour line appears between the dark and illuminated halves. The dark line is adjusted exactly on the cross wires and the refractive index is read on the scale.

Usually commercial instruments are constructed for use with white light but are calibrated to give the refractive index in terms of sodium light of wavelength, 589.3 nm at a temperature of 40°C unless otherwise specified (British Pharmacopoeia Appendix IV - B, p. 1016 (1963)).

Refractive Index

Definition

The refractive index of a substance is the ratio of the speed of light in a vacuum to the speed of light in the substance. The index of refraction of oils is characteristics within certain limits for each kind of oil. It is related to the degree of saturation but it is affected by other factors such as free fatty acid content, oxidation, and heat treatment.

Scope

Applicable to all normal oils and liquid fats.

Apparatus

Refractometer, any standard refractometer equipped with the Abbe Butyro or any standard scale. A dipping type refractometer is satisfactory. The temperature of the refractometer must be controlled within + / - 0.1°C and for this purpose it is preferably provided with a thermostatically controlled water bath and a motor driven pump to circulate water through the instrument. The instrument is standardised, following the manufacturer's instructions, with a liquid of known purity and refractive index. Distilled water, which has a refractive index of 1.3330 at 20°C is satisfactory in some cases.

Light source, if the refractometer is equipped with a compensator, a tungsten lamp or daylight bulb is permissible. Otherwise a monochromatic light such as an electric sodium vapour lamp is required.

Reagent

Toluene is satisfactory for cleaning the prisms lens. Tissue paper or cotton is also recommended for cleaning the prisms in order to avoid injury to them.

Procedure

Melt the sample, if it is not already liquid, and filter through filter paper to remove any impurities and the last traces of moisture. The sample must be completely dry.

The temperature of the refractometer is adjusted to 40°C for ordinary fats and oils. For higher melting point samples, use a temperature of 60°C.

Be sure that the prisms are clean and completely dry and then place several drops of the sample on the lower prism. Close the prisms and tighten firmly with the screw-head. Allow to stand for 1-2 minutes or until the sample comes to the temperature of the instrument.

Adjust the instrument and light to obtain the most distinct reading possible and then determine the refractive index. Take several readings and calculate the average.

Notes

Approximate temperature corrections may be made by the following calculations:

a. Butyro refractometer:

$$R = R' + K (T'-T)$$

R = the reading reduced to temperature T

R'= the reading at T ' °C

T = the standard temperature

T'= the temperature at which the reading R' is made

K= 0.55 for fats and 0.58 for oils.

b. Abbe's refractometer

The formula is identical with that given for Butyro's refractometer but K becomes 0.000365 for fats and 0.000385 for oils (AOAC, 1975).

10.3 ESTIMATION OF ACID VALUE

Fatty acids present in substantial amounts in fats of meat animals are oleic, palmitic and stearic. Free fatty acids are the products of enzymatic or microbial lipolysis of lipids and hence goes on increasing during storage of meat and is responsible for flavour determination in meat which is particularly noticeable when the acidity of the extracted fat reaches 2 to 3 percent (as oleic acid)

The acid value of a fat is the number of mg of potassium hydroxide (KOH) required to neutralise the free fatty acids in 1 g of the substance.

Reagents

1. A mixture of equal volume of alcohol (95percent) and ether.
2. 1 percent Phenolphthalein in alcohol.
3. 0.1 N KOH.

Procedure

Weigh about 10 g of the oil or fat, accurately into a 250 ml conical flask and add 50 ml of a mixture of equal volumes of alcohol and ether previously neutralised after the addition of 1 ml of phenolphthalein solution. If necessary, the contents may be warmed in a water bath until the substance has completely dissolved. Titrate the solution with N/10 KOH with constant shaking until a pink colour persists for 15 sec. The titre value in ml (a) is noted.

$$\text{Acid value} = A \times 0.00561 \times 1000 \,/\, \text{Weight in g of substance}$$

If the normality of KOH is not exactly 0.1 N and if it is N then the above equation is multiplied by the factor N / 0.1

10.4 ESTIMATION OF SAPONIFICATION VALUE

Reagents

1. KOH solution: 35 to 40 g of KOH pellets are dissolved in 20 ml of water to which is added sufficient alcohol (95 percent) to make 1liter. The solution is allowed to stand overnight and the clear supernatant is used for the estimation.
2. 0.5 N HCl
3. 1 percent Phenolphthalein solution in 95percent alcohol.

Procedure

Weigh about 2 g of the substance accurately in a 250 ml round bottom flask. Add 25 ml of the alcoholic KOH solution, attach a reflux condenser and heat the flask over boiling water bath for 1 hour. While the solution is still hot, add 1 ml of phenolphthalein solution and titrate the excess alkali against 0.5 N HCl (a). Repeat the experiment without the oil or fat to obtain the blank value (b) Since 1 ml of 0.5 N HCl is equivalent to 0.02805 g of KOH, the following equation is used to calculate the saponification value.

$$\text{Saponification value} = \frac{(b - a) \times 0.02805 \times 1000}{\text{Weight in g of substance}}$$

If the HCl has a normality of 'N' then the above equation is multiplied by a factor of N/ 0.5

10.5 ESTIMATION OF IODINE VALUE (WIJS METHOD)

Iodine value is a measure of the unsaturation of fats and oils and is expressed in terms of the number of g of iodine absorbed per gram of sample (percent iodine absorbed).

Apparatus

Glass stoppered bottles, 500 ml.

Glass stoppered volumetric flask , 1000 ml calibrated.

Pipettes 25 ml and 20 ml.

Reagents

1. Glacial acetic acid , reagent grade. The permanganate test should be applied to be sure that this specification is met. (Test :- dilute 2 ml of the glacial acetic acid with 10 ml of distilled water and add 0.1 ml of 0.1 N $KMnO_4$. The pink colour must not be entirely discharged within 2 hours).
2. Potassium iodide, reagent grade
3. Chlorine ,99.8 percent
4. Carbon tetrachloride, reagent grade
5. Hydrochloric acid , reagent grade, specific gravity 1.19.
6. Soluble starch

Test for sensitivity:- Make a paste with 1 g of starch and a small amount of cold distilled water. Add, while stirring, 200 ml of boiling water. Place 5 ml of this solution in 100 ml of water and add 0.5 ml of 0.1N iodine solution. The deep blue colour produced must be discharged by 0.05 ml of 0.1 N sodium thiosulphate.

7. Potassium dichromate, reagent grade. The potassium dichromate is finely ground and dried to constant weight at ca 110°C before using.

8. Sodium thiosulphate ($Na_2S_2O_3.5H_2O$), reagent grade

9. Iodine, reagent grade

Solutions

1. Potassium iodide solution, 15 percent

2. Starch indicator solution, weigh 10 g starch and add 10 ml cold distilled water. Add to this 1 liter of boiling distilled water and stir rapidly and cool. Salicylic acid (1.25 g per liter may be added as preservative. Starch solution must be stored in refrigerator.

3. Standard potassium dichromate solution. 0.1 N, dissolve 4.9035 g potassium dichromate in 1000 ml of distilled water in volumetric flask and make the volume.

4. Sodium thiosulphate solution, 0.1N, dissolve 24.8 g of sodium thiosulphate in distilled water and dilute up to 1 liter.

5. Wijs solution, dissolve 13.0 g of iodine in 1 liter of glacial acetic acid. Heat gently if necessary. Cool and remove a small quantity (100-200 ml) and set aside in a cool place for future use. Pass dry chlorine gas into the iodine solution until the original titration is not quite doubled. A characteristics colour change takes place in the Wijs solution when desired amount of chlorine is added. This may be used to assist in judging the end-point. A convenient procedure is to add a small excess of chlorine and bring back to the desired titration by addition of some of the original iodine solution, which was taken out at the beginning. The original iodine solution and finished Wijs solution are both titrated with $Na_2S_2O_3$ solution.

Procedure

1. Melt the sample (do not heat beyond 10-15°C above the melting point of the sample. Filter through filter paper to remove impurities and traces of moisture. The sample must be absolutely dry.

2. Weigh the sample accurately in 500 ml bottle or flask (stoppered). The weight of the sample must be such that there will be an excess of Wijs solution of 50 to 60 percent of the amount added i.e. 100 to150 percent of the amount absorbed. For meat fat, which has iodine value in the range of 45 to 65, the sample weight is generally about 5 g.

3. Add 20 ml of carbontetrachloride (CCl_4)

4. Pipette 25 ml Wijs solution into the flask containing sample and swirl to ensure proper mixing.

5. Prepare and conduct at least 2 blank determinations with each group of samples simultaneously in all respect.

6. Store flasks in dark place for 30 minutes at room temperature (25°C)

7. Remove the flasks and add 20 ml potassium iodide solution followed by 100 ml of distilled water.

8. Titrate with 0.1 N sodium thiosulphate adding gradually and vigorous shaking. Continue titration until the yellow colour has almost disappeared. Add 1 to 2 ml of starch indicator and continue titration until the blue colour has just appeared.

Calculation

$$\text{Iodine value} = (B\text{-}S) \times N \times 12.69 \ / \ \text{weight of the sample}$$

B = titration of blank,

S = titration of sample

N = normality of sodium thiosulphate solution

10.6 ESTIMATION OF IODINE VALUE (HANUS IODINE VALUE METHOD)

Principle

The unsaturated glycerides of fat have the ability to absorb a definite amount of iodine especially when aided by a carrier such as iodine chloride or iodine bromide and thus forms saturated compounds. The quantity of iodine absorbed is a measure of the un-saturation of fat and is expressed as the number of g of iodine absorbed by 100 g of fat or oil.

Reagents

1. Iodine bromine solution: The reagent used is Ibr (iodine bromide) and is prepared by dissolving 13.615 g pure iodine in 825 ml of glacial acetic acid cool and add 3 ml of bromine solution.

2. Chloroform reagent AR

3. Potassium iodide 15 percent in volumetric flask with freshly boiled and cooled distilled water.

4. Sodium thiosulphate 0.1N ($Na_2S_2O_3$) accurately standardised.

5. Starch indicator for titration.

Procedure

Weigh 0.5 g of fat in iodine flask and add 10 ml of chloroform.

Shake until fat is dissolved.

Add 25 ml of Hanus iodine solution and allow to stand for 30 minutes with occasional shaking.

Add 10 ml of 15 percent KI and shake.

Add 100 ml of boiled and cooled distilled water.

Titrate with 0.1 $Na_2S_2O_3$. As yellow colour becomes faint add few drops of starch indicator. When blue colour disappears, it is considered as an end point.

Calculation

Iodine / 10 g of fat = A × N of $Na_2S_2O_3$ × 126.91/ weight of sample

A = difference of (ml of $Na_2S_2O_3$ required for blank – ml of $Na_2S_2O_3$ required for sample).

10.7 ESTIMATION OF PEROXIDE VALUE TEST FOR LARD AND TALLOW

Introduction

This method determines, in terms of milli-equivalents peroxide per 1000 g of sample, which oxidises Potassium iodide (KI) under the conditions of test. These are generally assumed to be peroxides or other similar products of fat oxidation.

Scope:- Applicable to all normal fats and oils including margarine. This method is highly empirical and any variation in procedure may result in variation of results.

Reagents

Acetic acid Chloroform solution: Mix 3 parts by volume of glacial acetic acid AR grade with 2 parts by volume of Chloroform U.S.P. grade.

Potassium Iodide solution: Saturated solution of KI in recently boiled and cooled distilled water. Make sure the solution remains saturated as indicated by the presence of un-dissolved crystals. Store in the dark. Test daily by adding 2 drops of starch solution to 0.5 ml of the KI solution in 30 ml of acetic acid - Chloroform solution. If a blue colour is formed which requires more than 1 drop of 0.1 N sodium thiosulphate solution to discharge, discard the Iodide solution and prepare a fresh one.

Sodium thiosulphate solution: 0.1 N accurately standardised. This may be prepared by accurately pipetting 100 ml of normal solution into 1000 ml volumetric flask and diluting to volume with recently boiled distilled water.

Starch indicator solution: 1.0 percent soluble starch in distilled water.

Procedure

Weigh 10 g of sample into a 250 ml glass stoppered conical flask and add in portions 10 percent alcohol in distilled water shaking vigorously and 30 ml of acetic acid chloroform solution. Swirl the flask until the sample is dissolved. Add 0.5 ml of saturated KI preferably using Mohr type measuring pipette. Allow the solution to stand with occasional shaking for exactly 1 min. and then add 30 ml of distilled water. Titrate with 0.1 N Sodium thiosulphate adding it gradually and with constant and vigorous shaking. Continue the titration until the yellow colour has almost disappeared. Add 0.5 ml of starch indicator solution. Continue the titration, shaking the flask vigorously near the end point, to liberate all the iodine from the chloroform layer. Add thiosulphate drop wise until the blue colour has just disappeared. Note:- If the titration is less than 0.5 ml repeat the determination using 0.01 N Sodium thiosulphate solution. Conduct a blank determination of the reagents daily. The blank titration must not exceed 0.1 ml of the 0.1 N sodium thiosulphate solution.

Calculation

Peroxide value as milli equivalents of peroxide per 1000 g of sample =

$$S \times N \times 1000 \ / \ \text{weight of sample.}$$

S = titration value of sample

N = Normality of Sodium thiosulphate solution

10.8 ESTIMATION OF UNSAPONIFIABLE MATTER

Definition

Unsaponifiable matter includes those substances frequently found dissolved in fats and oils, which cannot be saponified by the caustic alkalies but are soluble in the ordinary fat solvents. Included are the higher aliphatic alcohol, sterols, pigments, and hydrocarbons.

Apparatus

1. Erlenmeyer flask, 200 ml capacity with water -cooled condenser
2. Separating funnels, 500 ml capacity.

Reagents

1. Ethyl alcohol, 95 percent
2. Aqueous potassium hydroxide solution, 50 percent KOH
3. Petroleum ether
4. Sodium hydroxide 0.02N
5. Phenolphthalein indicator 1.0percent in 95 percent alcohol.

Procedure

Weigh 5 g of the sample in to 200 ml Erlenmeyer flask. Add 30 ml of 95 percent alcohol and 5 ml of 50percent KOH. Boil gently but steadily under a reflux condenser for 1 hr or until completely saponified. Complete saponification is essential.

Transfer to the separating funnel and complete the transfer with warm water and then cold distilled water until the total volume is about 80 ml. Wash the flask with a little petroleum ether and add to the separating funnel. Cool the contents to room temperature (30-35°C) and then add 50 ml of petroleum ether.

Shake vigorously for at least 1 min and allow to settle until both layers are clear. Collect the upper layer in a weighed flask and dry at 80°C. Find out the weight of the residue.

Calculation

Unsaponifiable matter, percent

$$= \frac{(\text{weight of the residue - weight of fatty acids}) \times 100}{\text{weight of the sample}}$$

10.9 CLOUD TEST

Definition

The cloud point is that temperature at which, under the conditions of this test, a cloud is introduced in the sample caused by the first stage of crystallisation.

Apparatus

Oil sample bottle (115 ml), Thermometer

Water bath of ice water not less than 2°C or more than 5°C.

Procedure

The sample must be completely dry before making the test. Heat 60 to 75 ml sample to 130°C immediately before making test. Pour 45 ml of the heated fat into an oil sample bottle.

Begin to cool the bottle and contents in the water bath, stirring enough to keep the temperature uniform. When the sample has reached a temperature of 10°C above the cloud point, begin stirring steadily and rapidly in circular motion so as to prevent supercooling and solidification of fat crystals on the sides or bottom of the bottle.

From this point on do not remove the thermometer from the sample, since doing so may introduce air bubbles, which will interfere with the test. The test bottle is maintained in such a position that the upper levels of the sample in the bottle and the water in the bath are at the same level.

Remove the bottle from the bath and inspect regularly. The cloud point is that temperature at which that portion of the thermometer immersed in the oil is no longer visible when viewed horizontally through the bottle and sample.

10.10 TITER TEST

Definition: This method determines the solidification point of the fatty acids.

Apparatus

Griffin low form beaker, 2 liter capacity

Wide mouth bottle, capacity 450 ml, height 190 mm inside diameter 38 mm.

Test tubes, length 100 mm, diameter 25 mm with or without rim. These test tubes may have an etched mark extending around the tube at a distance of 57 mm from the bottom to show the height to which the tube is to be filled.

Procedure

Preparation of fatty acids:- Weigh 110 g of glycerol caustic in the saponification vessel. Stir and heat to 150°C. Add 50 ml. of oil or melted fat sample and reheat to 140°C to 150°C. In some cases a little additional solution may be necessary to insure complete saponification. Continue stirring until saponification is complete. Do not heat above 150°C. Cool slightly, add 200 to 300 ml of distilled water, stir the mass well and heat until the soap is dissolved. Add carefully while stirring 50 ml of dilute sulphuric acid. Boil until the fatty acids are completely melted and clear. Additional water may be added before or during boiling if desired. If a gas burner is used for heating, the water level is kept high enough to prevent scorching on the side of the dish.

Remove the aqueous layer containing the sulphuric acid. This may be done with appropriate siphon or by any other convenient means. Again add water and boil for 2 to3 minutes or until the fatty acids are entirely melted and clear. Acids of high melting point fats are sometimes slow to melt and clear. Inspect the fatty acid layer while it is quiet to be sure all has melted.

Remove the water again and if, necessary, repeat the washing as directed above until the wash water is neutral to methyl orange indicator.

Transfer the fatty acids to a filter paper carefully so as to include any water. The filter paper may be supported on a small beaker without a funnel. The acids must remain completely melted until entirely filtered.

Heat the filtered acids on a hot plate to 130°C to remove traces of moisture and fill a titre test tube to a height of 57 mm from the bottom.

Solidification of the Fatty Acids

Fill the water bath to the designated level and adjust the temperature to 20°C for all samples having titre of 35°C or higher and 15 - 20°C below the titer point for all samples with titer below 35°C.

Place the test tube containing the fatty acids in the assembly as shown in the figure. Insert the titre thermometer to the immersion mark so that it will be equidistant from the sides of the tube.

Stir, with the stirring rod in a vertical manner at the rate of 100 complete up and down motions per minute. The stirrer moves through a vertical distance of 38 mm. The agitation is started while the temperature is at least 10°C above the titre point.

Stir at the directed rate until the temperature remains constant for 30 seconds, or begins to rise in less than a 30 seconds interval. Discontinue stirring immediately, remove the stirrer or raise it out of the sample, and observe the increase in temperature. The tire point is the highest temperature indicated by the thermometer during this rise. Duplicate determinations are usually expected to agree within 0.2°C.

References

A.O.A.C. Official method Cc 6-25

A.O.A.C. Official method Cc 7-25.

A.O.A.C. Official method Cd 1- 25

A.O.A.C. Official method Cd 3-25.

A.O.A.C. Official method Da 11- 42.

A.O.A.C. Official method Cc 12-41.

British Pharmacopoeia Appendix IV - B, p. 1016 (1963).

British Pharmacopoeia Appendix IV - E, p. 1017 (1963).

British Pharmacopoeia Appendix X - A, p. 1055 (1963).

British Pharmacopoeia Appendix X - C, p. 1057 (1963).

Church, P.N. and Wood, J.M. (1992) In the manual of manufacturing meat quality, Elsevier Applied Science Publishers Limited, Crown house, Linton road, Barking, Essex, IG11 8JU, England.

Koniecko, E.K.(1979) In Hand book for Meat Chemists, pp 68-69, Avery Publishing Group Inc. Wayne, New Jersey, USA.

Kowale, B.N., Kulkarni, V.V. and Kesava Rao, V. (2008) Methods in meat science. *Pub.* by Jaypee brothers medical publishers (P) Ltd., Ansari Road, Daryagunj, New Delhi.

■ ■ ■

TECHNIQUES FOR MICROBIOLOGICAL QUALITY EVALUATION OF MEAT

11.1 SAMPLING PLAN FOR MICROBIOLOGICAL EXAMINATION OF SAMPLES

Statistical sampling plan ensure that the tests, which are done, are sufficient to give good assessment of the microbial condition of the food. It is important in minimising the chances of rejecting an acceptable lot of food or accepting substandard lot.

The sample should be representative as closely as possible to the actual consignment with respect to its microbial flora.

The number of samples depends upon:

● The uniformity or homogeneity of the product

● Size of many particles

● Previous knowledge and experience of material

● Potential health hazard of the food

Usually for small consignment, sample size is the square root of the number of the containers and for bigger consignment it is sufficient percentage of the total material. The sample should be collected aseptically and analysed quickly without many holding periods or preserved at refrigeration under certain conditions.

Sampling plan

A sampling plan is an appropriate examination of a required number of sample units by specified methods. A plan can be two (n, c, m) or three class (n, c, m, M) plan.

n = the number of sample units from a lot that must be examined to satisfy a given sampling plan.

c = Maximum allowable number of sample units that may exceed the microbiological 'm', when this number is exceeded, the lot is rejected.

m = The maximum number of relevant bacteria per gram. Values above this level are marginally acceptable or unacceptable. In the presence / absence situation for two class plan. It is common to assign m = 0, for three class plan, m is usually some non zero value.

M = It is used to separate marginally acceptable quality from unacceptable quality. Food values at or above M in any sample are unacceptable relative to health hazard, sanitary indicators or spoilage potential (Table 11.1 and 11.2).

Table 11.1: ICMSF sampling plan and recommended microbiological limits

Product	Tests	Class plan	n	c	m	M
Raw chicken						
(fresh/frozen)	APC	3	5	3	5×10^5	10^7
Comminuted raw meat (frozen) and chilled carcass meat	APC	3	5	3	10^6	10^7
Roast beef	*Salmonella*	2	20	0	0	-
Frozen raw	*S.aureus*	3	5	2	10^3	10^4
crustaceans	*V. parahaemolyticus*	3	5	1	10^2	10^3
(crabs, lobsters	*Salmonella*	2	5	0	0	-
etc.)	APC	3	5	2	5×10^5	10^7
	E.coli	3	5	2	11	500
Mineral water	*Coliform*	2	5	0	0	-

International Commission on Microbiological Specifications for Foods of the International Association of Microbiological Society (ICMSF) recommended that general viable count at 35°C or at 20°C in case of chilled meats should be less than 10^7/gram and the *Salmonella* should not be detected in more than one sample out of five 25 gram samples.

Table 11.2: Sampling plan and microbiological limits for processed meats

Product	Tests	n	c	M	M
Dried meats and	*Staphylococcus aureus*	5	1	10^2	10^4
products	*Clostridium perfringens*	5	1	10^2	10^4
	Salmonella	10	0	0	0
Dehydrated poultry products	*Salmonella*	10	0	0	0

[Table Contd.

Contd. Table]

Product	Tests	n	c	M	M
Roast beef	Salmonella	20	0	0	0
Cooked uncured meats	Salmonella	20	0	0	0
Canned cured meats	Salmonella	10	0	0	0
Poultry raw	Salmonella	5	0	0	-
	APC	5	3	5×10^5	10^7
Cooked poultry	Staphylococcus aureus	5	1	10^3	10^4
Meat frozen	Salmonella	5	0	0	0
Cured and smoked meat products	Staphylococcus aureus	10	1	10^3	10^4

Sampling, method for meat and meat products

1. Swab method : Wrap absorbent cotton firmly on a wooden splinter, approximately ¾ inch long, applicator to form a swab of about 1/16-inch diameter. Moisten the swab in sterile saline (0.85 %) after sterilising in hot air oven at 160°C for 1 h.

2. Take 0.25 sq. inch skin surface at different places i.e. breast, thigh, back, under wings etc. and swab 4 to 5 times in each direction rotating the swab in the process. Break the applicator stick about ½ inch above the swab aseptically and place the swab in a test tube containing 5 ml of sterile saline. Shake the tube vigorously by rotating it in between the palms to disperse the cotton (stock solution). Make serial dilutions and pour 0.5 ml of the diluent into petridish containing media and incubate it at 37°C for 24, 48 and 72 h.

3. Trituration method: Weigh 10 g of meat from different parts of the carcass viz. breast, back, thigh, drumstick, wing, etc. in a sterilized petridish and cut into small pieces. Triturate the meat with 90 ml of sterile physiological saline in a sterilised pestle and mortar. Take 1 to 2 drops of triturated sample for plating.

11.2 ENUMERATION OF TOTAL VIABLE COUNT

There are two types of microbiological test methods commonly used in meat microbiology.

Reference tests for microbiological quality evaluation of food products

Microbiology - General guidance for the enumeration of microorganisms - Colony count technique at 30° C	IS 5402:2002/ISO:4833:1991 reaffirmed 2007
Microbiology - General guidance for the enumeration of coliforms: part 1 Colony count technique (first revision) or general guidance for estimation of coliforms: part 2 most probable number technique (first revision)	IS 5401 (Part 1): 2002/ISO 4832:1991 Reaffirmed 2007 OR IS 5401(Part 2): 2002/ISO 4831:1991 reaffirmed 2007
Methods for detection of bacteria responsible for food poisoning: part 1 isolation, identification and enumeration of *E.coli*	IS 5887 (Part 1): 1976 reaffirmed 2009
Methods for detection of bacteria responsible for food poisoning: part 3 general guidance on methods for detection of *Salmonella*	IS 5887 (Part 3):1999/ ISO 6579:1993 reaffirmed 2009
Methods for detection of bacteria responsible for food poisoning: part 8 horizontal method for enumeration of coagulase-positive *Staphylococci* section 1 technique using Baird-Parker agar medium section 2 technique using rabbit plasma fibrinogen Agar medium	IS 5887 (Part 8/Sec 1): 2002/ISO 6888-1: 1999 reaffirmed 2007 OR IS 5887 (Part 8/Sec 2): 2002/ISO 6888-2: 1999 reaffirmed 2007
Method for yeast and mould count of food stuffs and animal feeds	IS 5403:1999 reaffirmed 2005/ISO 7954:1987 reaffirmed 2009
Indian standard specification for sterilized milk	IS: 4238-1967 Reaffirmed 2010
Methods for detection of bacteria responsible for food poisoning: part 6 identification, enumeration and confirmation of *B. cereus*	IS 5887(Part 6): 1999/ISO 7932:1993 reaffirmed 2007
Methods for detection of bacteria responsible for food poisoning: part 4 isolation, identification of *Clostridium perfringens, C. botulinum* and enumeration of *C. perfringens*	IS:5887 PART IV:1999 reaffirmed 2009
Microbiology of food and animal feeding stuffs - horizontal method for detection and enumeration of *Listeria monocytogenes* : part 1 detection method or part- 2 enumeration method	IS 14988 (Part 1): 2001 reaffirmed 2007/ISO 11290-1 :1996 OR IS: 14988 (Part 2): 2002 reaffirmed 2007/ ISO: 11290-2: 1998
Methods of sampling for milk and milk products	IS 11546:1999/ISO 707: 1997 reaffirmed 2010

1. **Total counts** – In this test, the number of organisms in a sample or from a swabbed area of surface is determined. This type of test is used for determining general microbiological status and for monitoring surfaces and machinery.

2. **Presence or absence test** – This test is only required to determine if a particular type of organism – usually a pathogenic e.g. *Salmonella* is present. Microorganisms of pathogenic significance, if present are expected to be so in very low numbers e.g., *Salmonella* absent in 25 g of sample.

Four basic methods employed for total bacterial count are:

- Standard plate count
- Most probable number (MPN) method as a statistical determination of viable cells
- Dye reduction technique to estimate numbers of viable cells that possess reducing capacities
- Direct microscopic counts (DMC) for both viable and nonviable cells.

Total viable count (TVC) or standard plate count (SPC) is used to provide an estimate of the total viable bacteria in a particular meat sample under aerobic / anaerobic condition or at mesophilic or psychrophilic temperature of growth. The TVC is useful in indicating overall microbiological quality of meat however for specific organisms differential count is more relevant.

Principle

Ten fold dilutions of the samples are made by using a suitable diluent (0.1% peptone water / normal saline solution) and appropriate dilutions are poured on plain agar plates and incubated at 37°C for 48 h. Colonies between 25 and 300 are counted and multiplied by the inverse of the dilution factor to get total viable count. The TVC is expressed as count per gram of sample.

Materials Required

Sterile petri dishes, test tubes, automatic graduated pipettes, blender, fine weighing balance, incubator, laminar flow, colony counter, refrigerator, bunsen burner, glass rod, 0.1 % peptone water / normal saline solution, plain agar.

Sampling

Meat sample (50 g) can be incised aseptically from carcass or the surface swabs from a defined area (one sq. inch) can be collected depending on the objective.

Procedure

Weigh 10 g of meat sample and transfer it in a blending vessel, add 90 ml of 1 % peptone / normal saline solution and blend at about 10,000 rpm for 2 min. This is 10^{-1} dilution.

Transfer 1 ml of this dilution (10^{-1}) to another tube containing 9 ml normal saline solution. Like wise make serial ten fold dilutions up to 10^{-9} and discard 1 ml from the last tube.

One ml of the desired dilutions (10^{-5}, 10^{-6}, 10^{-7}) is poured on the already prepared plate count agar plates in triplicate. Spread by rotating the plates carefully and then incubate them in an incubator pre / set at 37°C for 48 h. Incubate one set of plates at 4°C in the BOD incubator to calculate the TVC of the psychrophilic bacteria.

Remove the plates and count the colonies between 30 and 300. Take the average and multiply the reciprocal of the dilution factor with the number of colonies. This gives the number of living organisms present in the sample.

Express the TVC as the total number of organisms (colony forming units, CFU) per gram of the sample.

11.3 ENUMERATION OF DIFFERENTIAL COUNT IN MEAT SAMPLE

Differential count refers to an estimate of the different types of organisms including fungi and yeast present in meat sample. Differential counts are generally estimated to know the presence of certain pathogenic organisms such as *Salmonella, E. coli, Staphylococci,* and *Clostridia.* These organisms grow on specific culture media and some may also require prior enrichment. The limits for the presence or absence of these pathogenic organisms in foods are different. Hence great care must be taken in estimating the counts.

Procedure

The serial ten fold dilutions are made as mentioned under TVC and the appropriate dilutions are streaked on the selective culture media with the help of platinum loop and the plates are incubated at respective temperatures as given in the Table 11.3.

(a) Yeast and mould count: Take 0.5 ml of serial dilutions as prepared above into petri dishes, add 10-15 ml of PDA (Potato dextrose agar) media, mix gently, let the media solidify and incubate the petri dishes at 32 ± 0.5°C. Count the number of yeasts and moulds after 4 days of incubation.

Table 11.3: Cultural characteristics of some microorganisms

S. no.	Organism	Gram reaction	Culture media	Growth Temp.	Colony type
1	*Salmonella* sp.	Gram negative rods, aerobic, motile, non-spore forming	MacConkey agar Brilliant green agar (BGA)	37°C	Pale, non lactose fermenting, smooth, small Bright pink colonies
			Xylose lysine desoxycholate agar (XLD)		Red colonies with black centre due to H_2S production
			Hektoen enteric agar (HE agar)		Green / blue green colonies with black centre due to H_2S production.
2	*Shigella*	Gram negative rods aerobic	Violet red bile agar (VRB)	37°C	Circular, large moist and grey colonies
3	*Staphylococcus aureus*	Gram positive cocci, clusters like bunch of grapes, aerobic	Nutrient agar	37°C	Circular, convex, glistening, having entire edges and pigmented and white to lemon yellow in colour
			Blood agar (sheep / cattle blood)		α, β, δ haemolysis
			Mannitol salt phenol red agar		Yellow colonies due to fermentation of mannitol
			Baird–Parker medium		Black colonies, round convex
			Carters medium		Fried egg type colonies
4	Streptococci	Gram positive cocci, chain, aerobic, haemolytic	Media containing blood	37°C	Small dew drop like colonies
			Blood agar		α, β, δ haemolysis
			Edward medium		Small transparent, bluish grey colour colonies, *Streptococcus agalactiae* produce green colonies.

[Table Contd.

Contd. Table]

S. no.	Organism	Gram reaction	Culture media	Growth Temp.	Colony type
5	*Bacillus anthracis*	Gram positive rods, aerobic, spore former	Nutrient agar	37°C, 48 h	Dull surface colonies of moderate size and grayish-white (frosted glass) appearance, on magnifying the colonies resemble locks of hairs or give an appearance of medusa head.
			Blood agar		No haemolysis
6	*Clostridium perfringens* Grows in anaerobic jars with atmosphere of 90 percent hydrogen and 10 percent carbon dioxide or 80 percent nitrogen, 10 percent hydrogen and 10 percent carbon dioxide after evacuation of air.	Gram positive rods, anaerobic, sporeformers	Sodium polymixin sulpha diazine medium (SPS)	37°C anaerobic	Black, cotton wool like colonies,
			Robertson's cooked meat medium		Pinkish / blackish colouration of the medium due to saccharolytic / proteolytic organisms of *Clostridium* group.
			Blood agar, anaerobic broth, egg yolk agar		Colonies are diffuse, greyish, semi-transparent and irregular in shape.
7	*Corynebacterium*	Gram positive small pleomorphic rods with Chinese letter arrangements, aerobic, facultatively anaerobic	Ordinary nutrient media	37°C	Round smooth colonies
			Blood agar		Greyish in colour about 1 mm in diameter surrounded by zone of haemolysis

[Table Contd.

Contd. Table]

S. no	Organism	Gram reaction	Culture media	Growth Temp.	Colony type
8	Escherichia coli	Gram negative rods or cocobacilli, aerobic, facultatively anaerobic	MacConkey medium	37°C	Pink dry colonies, lactose positive areas of precipitated bile salts.
			Eosin methylene blue agar (EMB)		Metallic sheen colonies
			Brilliant green agar (BGA)		Yellowish green colonies
			Xylose lysine desoxycholate agar (XLD)		Yellow colonies
9	Pseudomonas	Gram negative medium sized rods	Ordinary culture medium	37°C	Irregular colonies, spreading, translucent and 3-5 mm in diameter and may show metallic sheen,
			Blood agar		Beta haemolysis around the colonies, distinctive grape like odour is usually apparent whichever medium is used for culture.
10	Micrococci	Gram negative cocci	Mannitol salt agar (MSA)	37°C	Small dew drops like colonies, faint blue in colour.
11	Shigella sp.	Gram negative	MacConkey Agar	37°C	Clear, colourless colonies
			HE agar		Green colonies without black centre
12	Bacillus cereus	Gram positive motile, spore forming aerobic large rods	Blood agar	37°C	Frosted glass colonies

[Table Contd.

Contd. Table]

S. no.	Organism	Gram reaction	Culture media	Growth Temp.	Colony type
13	Fungi (*Aspergillus*)		Sabouraud's agar or Potato dextrose agar	37°C for 3-4 days	White filamentous growth, 2-3 mm above the surface and becomes green or dark green in colour.
14	Yeast (*Candida*)		Sabouraud's glucose agar or Potato dextrose agar	37°C for 3-4 days	Small, smooth and convex with creamy colour having yeast like odour.

(b) Coliform count: Do the plating as mentioned above except that use VRBA (Violet red bile agar) media and incubate the plates at $37 \pm 0.5°C$. Count the number after 24 hours of incubation.

(c) *Salmonella:* Take 25 g of sample and transfer directly to tetrathionate broth in a lot and shake for 1 minute. Incubate the flask $37 \pm 0.5°C$ for 24 hours. Streak a loop full from the enrichment broth on 'Brilliant green agar' and incubate the plates at $37 \pm 0.5°C$. Examine the plates after 24 and 48 hours of incubations. Appearance of pink or red colour colonies indicates presence of *Salmonella*.

11.4 ISOLATION OF *Listeria*

There are several methods for the isolation of *Listeria* spp. Many laboratories follow methods designed by Food and Drug Administration (FDA), International Organization for Standardization (ISO) and United States Department of Agriculture (USDA). The FDA method has been designed for the processing dairy products whereas USDA method is recommended for meat and poultry products. In the **FDA method** the sample (25 g is enriched) for 48 h at 30°C in *Listeria enrichment* broth (LEB) containing selective agent acriflavin and nalidixic acid and the antifungal agent cycloheximide, followed by plating into selective agar (Oxford, PALCAM or LPM).

The **ISO 11290 method** employs a two stage enrichment process: the first enrichment in half Fraser broth for 24 h then an aliquot is transferred to full-strength Fraser broth for further enrichment followed by plating on Oxford and PALCAM agars. Fraser broth also contains the selective agent acriflavin and nalidixic acid as well as esculin, which allows detection of β-D-glucosidase activity by *Listeria*, observed as blackening of the growth.

The **USDA method** contains modification of University of Vermont media (UVM) containing acriflavin and nalidixic acid for primary enrichment followed by secondary enrichment in Fraser broth and plating onto Modified Oxford (MOX) agar containing the selective agents moxalactam and colistin sulphate.

11.5 MOLECULAR DETECTION OF *L. monocytogenes* BY PCR

PCR is very rapid, easy, reliable, reproducible and quick technique for detection of any gene and therefore the organism. Traditional method of *L. monocytogenes* needs technical skill, laborious and time consuming. Here, we demonstrate the detection of *L. monocytogenes* based on three virulence associated genes

The protocol for isolation and identification of Listeria monocytogenes from clinical cases and foods

ISOLATION
(Clinical samples, food)
↓

SELECTIVE ENRICHMENT
USDA (University of Vermont Medium I and II) or Fraser broth
↓

SELECTIVE PLATING
PALCAM Agar
↓

CONFIRMATION
Routinely used methods
↓

PHENOTYPE		**GENOTYPE**
CULTURE BIOCHEMISTRY		PCR
Gram Stain		Sugar Fermentation
Motility		Biochemical testing
	Hemolysis	
	Camp Test	
	PI-PLC Assay	
	TYPING	
	Serotyping	
	RAPD	

Typical Isolation of *L. monocytogenes* by USDA method

Listeria **colonies on PALCAM agar**

exclusively present in the *L. monocytogenes* i.e. gene responsible for haemolysin (*hlyA*), phosphatidylinositol-phospholipase C (*plcA*) and listerial actin-polymerizing protein (*actA*).

Requirement

a) *Taq* DNA Polymerase,

b) 10X PCR buffer,

c) MgCl-$_2$,

d) dNTPs,

e) DNA template,

f) Eppendorf tubes,

g) PCR tubes,

h) Deionized water and

i) Specific Primers (Table)

Target Gene	Primer sequences (5'-3')	Product Size (bp)
hlyA	5'GCA GTT GCA AGC GCT TGG AGT GAA3' 5'GCA ACG TAT CCT CCA GAG TGA TCG3'	456
plcA	5'CTG CTT GAG CGT TCA TGT CTC ATC CCC C3' 5'CAT GGG TTT CAC TCT CCT TCT AC3'	1484
actA	5'CGC CGC GGA AAT TAA AAA AAG A3' 5'ACG AAG GAA CCG GGC TGC TAG3'	839
prfA	5'ACA AGC TGC ACC TGT TGC AG3' 3'TGA CAG CGT GTG TAG TAG CA5'	1060
iapA	5'ACA AGC TGC ACC TGT TGC AG3' 5'TGA CAG CGT GTG TAG TAG CA3'	131

Protocol

DNA preparation:

1. Transfer a single colony to the 1.5 ml eppendorf tube containing 100 μl of TE buffer.

2. Add 5 μl of lysozyme (10 mg/ml) and incubate at 37°C for 5 min. with gentle shaking.

3. Incubate the tube in boiling water for 10 min.

4. Centrifuge at 10000 rpm for 5 min.

5. Pipette the clear supernatant as a crude DNA to the new eppendorf tube.

6. Use 2 µl of the above supernatant as DNA template

PCR Protocol:

1. Prepare the reaction mixture as below:

Sr. No.	Name	Add (µl) per rct.	Final Conc.
1	10x PCR buffer	2.5	1X
2	MgCl₂ (25mM)	3.0	3 mM
3	dNTP mix (10mM)	2.0	0.8mM
4	hlyA (100µM) (F&R)	0.25 each	1µM
5	plc A (100µM) (F&R)	0.25 each	1µM
6	act A (100µM) (F&R)	0.25 each	1µM
7	*Taq* DNA Polymerase (5U/µl)	0.5	2.5 U per reaction
8	D/W	16.5	—
	Total	**25.0**	

2. Adjust the Reaction condition at your thermal cycler as below:

Temp. (°C)	Time (min)	Cycle
94	5	1
94	30 sec	30
60	30 sec	
72	45 sec	
72	5 min	1
4	Hold	

3. Place your tubes in the thermal cycler and put on the machine.

4. After respected time run the electrophoresis along with 100bp DNA marker and observe the respected bands.

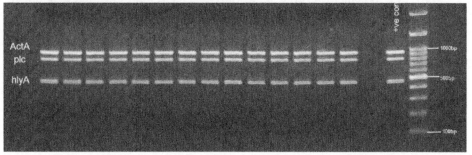

Fig. 1. Amplification of *act*A (965bp), *plc*A (803bp) and *hly*A (456bp) gene from the *Listeria monocytogenes* isolates

11.6 MULTIPLEX PCR BASED SEROTYPING OF *L. monocytogenes*

Serotyping is often used as a first-line typing method. Serotyping is widely used for long-term microbiological surveillance of human listeriosis. Although 13 serovars are described for the species *L. monocytogenes*, at least 95% of the strains isolated from foods and patients are of serovars 1/2a, 1/2b, 1/2c, and 4b. The serovar information allows discrimination between isolates probably belonging to an outbreak and those that are not part of the outbreak and thus decreases the number of strains which need to be further characterized. For the food industry, where the presence of *L. monocytogenes* is a big concern, tracing contaminating strains within the food chain and the plant environment is of primary importance. Therefore, the method proposed here should provide more powerful, as the targeting of different marker genes. Proposed multiplex PCR assay is highly specific for pathogenic *L. monocytogenes*, providing thus in the same step a species confirmation also. This multiplex PCR assay to separate the four major *L. monocytogenes* serovars (1/2a, 1/2b, 1/2c, and 4b). The PCR test which constitutes a rapid and practical alternative to laborious classical serotyping.

Requirement

a) *Taq* DNA polymerase,

b) 10X PCR buffer,

c) $MgCl_2$,

d) dNTPs,

e) DNA template,

f) Eppendorf tubes,

g) PCR tubes,

h) Deionized water and

i) Specific primers (Table)

   Taq DNA polymerase, 10X PCR buffer, $MgCl_2$, dNTPs, DNA template, eppendorf tubes, PCR tubes, deionized water and primers (Table)

Gene target	Primer sequences (5'-3')	Product Size (bp)	Serovar specificity
lmo0737	5'AGG GCT TCA AGG ACT TAC CC3' 5'ACG ATT TCT GCT TGC CAT TC3'	691	*L. monocytogenes* serovars *1/2a,1/2c,3a,3c*
lmo1118	5'AGG GGT CTT AAA TCC TGG AA3' 5'CGG CTT GTT CGG CAT ACT TA3'	906	*L. monocytogenes* serovars *1/2c, 3c*

[Table Contd.

Contd. Table]

Gene target	Primer sequences (5'-3')	Product Size (bp)	Serovar specificity
ORF2110	5'AGT GGA CAA TTG ATT GGT GAA3' 5'CAT CCA TCC CTT ACT TTG GAC3'	597	*L. monocytogenes* serovars *4b, 4d,* and *4e*
ORF2819	5'AGC AAA ATG CCA AAA CTC GT3' 5'CAT CAC TAA AGC CTC CCA TTG3'	471	*L. monocytogenes* serovars *1/2b, 3b, 4b, 4d and 4e*
prs	5'GCT GAA GAG ATT GCG AAA GAA G3' 5'CAA AGA AAC CTT GGA TTT GCG G3'	370	All *Listeria* spp.

Protocol

DNA preparation:

1. Transfer a single colony to the 1.5 ml eppendorf tube containing 100 μl of TE buffer.

2. Add 5 μl of lysozyme (10mg/ml) and incubate at 37°C for 5 min. with gentle shaking.

3. Incubate the tube in boiling water for 10 min.

4. Centrifuge at 10000 rpm for 5 min.

5. Pipette the clear supernatant as a crude DNA to the new eppendorf tube. Use 2 μl of the above supernatant as DNA template

PCR Protocol:

6. Prepare the reaction mixture as below:

Reaction Mixture

Sr. No.	Name	Add (μl) per rct.	Final Conc.
1	10x PCR buffer	2.5	1X
2	MgCl$_2$ (25mM)	3.0	3 mM
3	dNTP mix (10 mM)	2.0	0.8mM
4	*lmo0737* (100 μM)	0.25	1μM
5	*lmo1118* (100 μM)	0.37	1.5μM
6	*ORF2110* (100 μM)	0.25	1μM
7	*ORF2819* (100 μM)	0.25	1μM
8	*Prs* (100 μM)	0.05	0.2μM
9	Taq DNA polymerase (5 U/μl)	0.5	2.5 U /reaction
10	D/W	16.5	—
	Total	**25.0**	

7. Adjust the reaction condition at your thermal cycler as below:

Temp. (°C)	Time (min)	Cycle
94	5	1
94	5	1
94	30 sec	30
54	30 sec	
72	30 sec	
72	5 min	1

8. Place your tubes in the thermal cycler and put on the machine.
9. After respected time run the electrophoresis along with 100bp DNA marker and observe the respected bands.

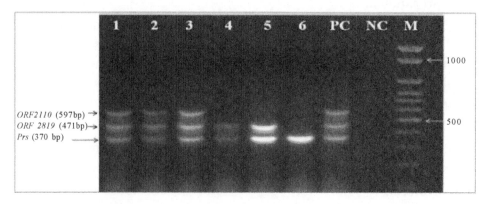

Fig. 2. Multiplex PCR serotyping for the determination of the serogroups of the isolates; Lane 1: Isolate FC2 *L. monocytogenes* serogroup 4b, 4d, 4c; Lane 2: 16 *L. monocytogenes* serogroup: 4b, 4d, 4c; Lane 3: 13 *L. monocytogenes* serogroup 4b, 4d, 4c; Lane 4: 22 *L. monocytogenes* serogroup 1/2b, 3b, 7; Lane 5:19 *L. monocytogenes* serogroup 1/2b, 3b, 7; Lane 6: *Listeria* spp., Positive control (PC); *L. monocytogenes* MTCC 1143; NC: Negative control; M: 100 bp DNA ladder.

11.7 DETECTION AND CONFIRMATION OF *Coliforms* AND *Escherichia coli*

Culture Media

*   Buffered Peptone Water (diluent)
*   Violet Red Bile Agar (VRBA)
*   Lauryl Sulfate Tryptose Broth (LST Broth)
*   Brilliant Green Lactose Bile Broth (BGLB broth)

- EC Broth (*Escherichia coli* Broth)
- MacConkey Agar
- Eosin Methylene Blue Agar (EMB Agar)
- Tryptone Water Broth
- MRVP Broth (Methyl Red-Voges-Proskauer Broth)
- Simmons' Citrate Agar (Citrate Agar)

Procedure

Test for Coliforms

Coliforms in foods may be enumerated by the solid medium method or by the Most Probable Number (MPN) method.

Solid medium method

Preparation of food homogenate: Make a 1:10 dilution of the well mixed sample, by aseptically transferring sample to the desired volume of diluent. Measure non-viscous liquid samples (i.e., viscosity not greater than milk) volumetrically and mix thoroughly with the appropriate volume of diluent (1 ml into 99 ml, or 10 ml into 90 ml or 50 ml into 450 ml).

- Weigh viscous liquid sample and mix thoroughly with the appropriate volume of diluent (1g into 99 ml; 10 g into 90 ml or 50 g into 450 ml). Weigh 50 g of solid or semi-solid sample into a sterile blender jar or into a stomacher bag. Add 450 ml of diluent. Blend for 2 minutes at low speed (approximately 8000 rpm) or mix in the stomacher for 30-60 seconds.

- Powdered samples may be weighed and directly mixed with the diluent. Shake vigorously (50 times through 30 cm arc). In most of the food samples particulate matter floats in the dilution water. In such cases allow the particles to settle for two to three minutes and then draw the diluent from that portion of dilution where food particles are minimum and proceed.

Dilution: If the count is expected to be more than 2.5×10^3 per ml or g, prepare decimal dilutions as follows. Shake each dilution 25 times in 30 cm arc. For each dilution use fresh sterile pipette. Alternately use auto pipette. Pipette 1 ml of food homogenate into a tube containing 9 ml of the diluent. From the first dilution transfer 1 ml to second dilution tube containing 9 ml of the diluent. Repeat using a third, fourth or more tubes until the desired dilution is obtained.

Pour Plating: Pipette 1 ml of the food homogenate (prepared sample) from each dilution into each of the appropriately marked duplicate petri dishes. In each petri-dish, pour 10-12 ml of violet red bile agar (VRBA) (cooled to 48°C) and swirl plates to mix. Allow it to solidify. Overlay with 3 to 5 ml VRBA and allow it to solidify. Incubate the dishes, inverted at 35°C for 18 to 24 hours.

Counting the colonies: Following incubation, count all colonies that are purple red in colour, 0.5 mm in diameter or larger and are surrounded by a zone of precipitated bile acids. Optimally the plates should have 30 to 100 colonies.

Calculation: Multiply the total number of colonies per plate with the reciprocal of the dilution used and report as coliforms per g or ml.

Most probable number method

This method is valuable in those samples where coliform density is low because higher quantity of sample can be used for examination. It is based on probability statistics wherein aliquots of decimal volumes/dilutions of the sample are transferred to several (1 to 5) tubes of specific medium. Positive tubes are scored and the MPN estimate is directly made using the Table 11.4.

Food homogenate is prepared. Preparation and dilutions were made as mentioned earlier.

Inoculation: Inoculate each of 3 tubes of Lauryl sulphate tryptose broth (LST broth) (containing inverted Durham tubes) with 1ml of food homogenate (1:10). Carry out the same operation from the first (1 in 100) and the second (1 in 1000) dilution tubes. Using a fresh sterile pipette for each dilution. Incubate the LST tubes at 35°C for 24 and 48 hours.

Presumptive test for coliforms: Record tubes showing gas production after 24 hours and re-incubate negative tubes for further 24 hours. Then record tubes showing gas production.

Confirmed test for coliforms: Transfer a loopful from each gas positive tube of LST to a separate tube of brilliant green lactose bile (BGLB) broth. Incubate the BGLB broth tubes at 35°C for 48h. The formation of gas confirms the presence of coliform bacteria. Record the number of positive tubes that were confirmed as positive for coliform.

Calculation: Note the MPN appropriate to the number of positive tubes from the table 11.4.

Table 11.4: Most probable number index to enumerate coliform count.

Positive tubes				Positive Tubes				Positive tubes				Positive tubes			
0.1	0.01	0.001	MPN	0.1	0.01	0.001	MPN	0.1	0.01	0.001	MPN	0.1	0.01	0.001	MPN
0	0	0	<3	1	0	0	3.6	2	0	0	9.1	3	0	0	23
0	0	1	3	1	0	1	7.2	2	0	1	14	3	0	1	39
0	0	2	6	1	0	2	11	2	0	2	20	3	0	2	64
0	0	3	9	1	0	3	15	2	0	3	26	3	0	3	95
0	1	0	3	1	1	0	7.3	2	1	0	15	3	1	0	43
0	1	1	6.1	1	1	1	11	2	1	1	20	3	1	1	75
0	1	2	9.2	1	1	2	15	2	1	2	27	3	1	2	120
0	1	3	12	1	1	3	19	2	1	3	34	3	1	3	160
0	2	0	6.2	1	2	0	11	2	2	0	21	3	2	0	93
0	2	1	9.3	1	2	1	15	2	2	1	28	3	2	1	150
0	2	2	12	1	2	2	20	2	2	2	35	3	2	2	210
0	2	3	16	1	2	3	24	2	2	3	42	3	2	3	290
0	3	0	9.4	1	3	0	16	2	3	0	29	3	3	0	240
0	3	1	13	1	3	1	20	2	3	1	36	3	3	1	460
0	3	2	16	1	3	2	24	2	3	2	44	3	3	2	1100
0	3	3	19	1	3	3	29	2	3	3	53	3	3	3	>1100

For example: 3 in 1:10; 1 in 1:100 and 0 in 1:1000. The table shows that MPN = 43 coliforms per g or ml.

Coliforms = present/absent per g

Test for faecal coliforms: Transfer a loopful from each gas positive tube of LST to a separate tube of EC broth. Incubate the EC tubes at 45°C in water bath for 24 hours. Submerge broth tubes so that water level is above the highest level of medium. Record tubes showing gas production.

Test for *Escherichia coli*: Streak one plate of MacConkey agar and EMB agar from each positive BGLB tube in a way to obtain discrete colonies. Incubate inverted plates at 37°C for 24 hours. On MacConkey agar, typical red to pink colonies will appear due to lactose fermentation. On EMB, examine plates for typical nucleated dark centered colonies with or without sheen. If typical colonies are present pick two from each EMB plate by touching needle to the center of the colony and transfer to a nutrient agar (NA) slant. Incubate slants at 37°C for 24 h.

Biochemical tests: Transfer growth from NA slants to the following broth for biochemical tests.

Tryptone broth: Incubate at 37°C for 24 hours and test for indole.

MR-VP Medium: Incubate at 37°C for 48 hours. Aseptically transfer 1 ml of culture to a 13x100 mm tube and perform the Voges-Proskauer test. Incubate the remainder of MR-VP culture an additional 48 h and test for methyl red reaction.

Simmons' citrate agar: Incubate 96 hours at 37°C and record as + or − for change in color to blue.

LST broth: Incubate 48 hours at 37°C and examine for gas formation.

Gram stain: Perform the Gram stain in a smear prepared from 18 hours NA slant. Presence of small red coloured rods confirms *Escherichia coli.*

Compute MPN of *E. coli* per g or ml considering gram negative, non-spore forming rods producing gas in lactose and classify biochemical types as follows (IMViC).

Indole test	MR test	VP test	Citrate test	Type of organism
+	+	-	-	Typical *E. coli*
-	+	-	-	Atypical *E. coli*
+	+	-	+	Typical intermediate
-	+	-	+	Atypical intermediate
-	-	+	+	Typical *Elnterobacter aerogenes*
+	-	+	+	Atypical *E. aerogenes*

Colonies of *E. coli* on EMB agar with metallic sheen.

11.8 DETECTION AND CONFIRMATION OF *Salmonella* SPECIES

Culture Media

⇨ Lactose Broth

⇨ Trypticase Soy Broth

⇨ Selenite Cystine Broth

⇨ Tetrathionate Broth

⇨ Xylose Lysine Deoxycholate (XLD) Agar

⇨ Hektoen Enteric Agar (HEA)

⇨ Bismuth Sulphite Agar (BSA)

⇨ Triple Sugar Iron (TSI) Agar

⇨ Lysine Iron Agar (LIA)

⇨ Urea Broth

⇨ Phenol Red Carbohydrate Broth

⇨ Tryptone Broth

⇨ Kcn Broth

⇨ Malonate Broth

⇨ Buffered Glucose (MR-VP) Medium

⇨ Brain Heart Infusion (BHI) Broth

⇨ Buffered Peptone Water (BPW)

Preparation of sample and pre-enrichment

Aseptically open the sample container and weigh 25 g of sample into a sterile empty wide mouth container with screw cap or suitable closure. Add 225 ml of sterile lactose broth/ buffered peptone water/ trypticase soy broth/ nutrient broth to the sample for pre- enrichment. Make a uniform suspension by blending if necessary. Cap container and let stand at room temperature for 60 min. Incubate at 37°C for 24 hours.

Selective enrichment: Gently shake incubated sample mixture and transfer 1 ml to 10 ml of selenite cysteine broth and an additional 1 ml to tetrathionate broth. Incubate for 24 hours at 37°C.

Selective media plating: Vortex – mix and streak 3 mm loopful of incubated selenite cystine broth on selective media plates of XLD, HEA and BSA. Repeat with 3 mm loopful of incubated tetrathionate broth. Incubate plates at 37°C for 24 and 48 hours. Observe plates for typical *Salmonella* colonies on XLD (after 24 h) - Pink colonies with or without black centres. On HEA (after 24 h) - Blue green to blue colonies with or without black centers. On BSA (after 24 to 48 h) – Brown, grey or black colonies sometimes with metallic sheen. Surrounding medium is usually brown at first, turning black with increasing incubation time.

Treatment of typical or suspicious colonies: Pick with needle typical or suspicious colonies (if present) from each XLD, HEA and BSA plates. Inoculate portion of each colony first into a TSI agar slant, streaking slant and stabbing butt and then do the same into an LIA slant. Incubate TSI and LIA slants at 37°C for 24 h and 48 h respectively. Cap tubes loosely to prevent excessive H_2S production.

Typical *Salmonella* reactions are:

	TSI	LIA
Slant	Alkaline (red)	Alkaline (Purple)
Butt	Acid (Yellow)	Alkaline (Purple)
H_2S production (blackening in butt)	+ or -	+

A culture is treated as presumptive positive if the reactions are typical on either or both TSI and LIA slants.

Biochemical tests: Using sterile needle inoculate a portion of the presumptive positive culture from TSI slant into the following broths. Incubate at 37°C for the specified period of days and read for *Salmonella* typical reactions.

Broth/ Media	Time of incubation	Results
Urea broth	24h	Negative (no change in yellow colour of medium)
Phenol red lactose broth	48h	Negative for gas and/or acid reaction

[Table Contd.

Contd. Table]

Broth/ Media	Time of incubation	Results
Phenol red sucrose broth	48h	Negative for gas and/or acid reaction
Phenol red dulcitol broth	48h	Postive for gas and/or acid reaction
Tryptone broth	24h	Negative for indole test
KCN broth	48h	Negative (no turbidity)
Malonate broth	48h	Negative (green colour unchanged)
MR-VP medium	48h	Negative for VP test but positive for MR test.

Criteria for discarding Non-*Salmonella* Cultures

Test(s) or Substrate(s)	Results
Urease test	Positive (purple-red)
Indole test	Positive (red)
Flagellar test (Polyvalent or spicer- Edwards)	Negative (no agglutination)
Lysine decarboxylase test	Negative (yellow)
KCN broth	Positive (growth)
Phenol red lactose broth	Positive (acid and/or gas)
Lysine decarboxylase test	Negative (yellow)
Phenol red sucrose broth	Positive (acid and/or gas)
Lysine decarboxylase test	Negative (yellow)
KCN broth	Positive (growth)
Voges-Proskauer test	Positive (red)
Methyl red test	Negative (yellow)

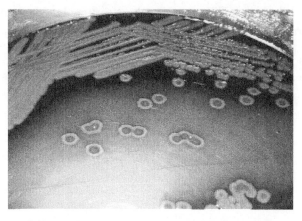

***Salmonella* colonies on Hektoen Enteric Agar**

11.9 DETECTION AND CONFIRMATION OF *SHIGELLA* SPECIES

Culture media

⇨ Gram Negative (GN) Broth

⇨ MacConkey Agar

⇨ Xylose Lysine Deoxycholate (XLD) Agar

⇨ Triple Sugar Iron (TSI) Agar Slants

⇨ Urea Broth

⇨ Acetate Agar Slants

⇨ Carbohydrate Fermentation Media

⇨ Tryptone Broth (for Indole Test)

⇨ Buffered Glucose (MR-VP) Medium

⇨ Simmon's Citrate Agar

⇨ Decarboxylase Test Media (with Lysine or Ornithine or Arginine)

⇨ Motility Test Medium

Enrichment: Using aseptic techniques mix or blend if necessary 25 g sample with 225 ml of gram negative broth. Transfer to a sterile 500 ml bottle. Adjust pH (if necessary) to 6.0 - 7.0 with sterile 1N NaOH or 1N HCl. Incubate at 37°C for 18 hours.

Selective streaking: Transfer a 5mm loopful of the enrichment broth culture to the surface of MacConkey agar and XLD agar plates and streak to obtain isolated colonies. Incubate plates at 37°C for 24 h. Typical *Shigella* colonies on XLD agar appear as red or pink colonies usually about 1mm in diameter and on MacConkey agar as opaque or transparent colonies. Inoculate each suspected colony into TSI agar slant by streaking the slant and stabbing the butt. After overnight incubation at 37°C, typical *Shigella* reaction is alkaline (red) slant and acid (yellow) butt with no H_2S or gas production.

Biochemical tests: Perform the following biochemical tests on a portion of the suspected *Shigella* culture from the TSI slant

Test	Reaction
Motility	-
Urease	-
Acetate utilization	-
Gas from glucose	-

[Table Contd.

Contd. Table]

Test	Reaction
IMViC reaction	+ + - - or - + - -
Lysine decarboxylase	-
Arginine dihydrolase	- or +
Ornithine decarboxylase	+ or -

11.10 DETECTION, AND CONFIRMATION OF *Staphylococcus aureus*

Culture media

- Tripticase (tryptic) soy broth with 10% sodium chloride and 1% sodium pyruvate.
- Baird parker (BP) medium
- Brain heart infusion (BHI) broth
- Desiccated coagulase plasma (rabbit) with EDTA
- Butterfields buffered phosphate diluent
- Plate count agar (PCA)

Procedure

Preparation of food homogenate and dilution

Aseptically weigh 50 g food sample into the sterile blender jar. Add 450 ml of diluent (1:10) and homogenize 2 min at high speed (16000-18000 rpm). Pipette 10 ml of the food homogenate into 90 ml of diluent (or 1 ml to 9 ml) to make a 1:100 dilution. Mix well using a vortex-mixer. Transfer 1ml from this dilution to a fresh tube of 9 ml to give a 1:1000 dilution.

Most probable number method: This procedure is recommended for testing processed foods likely to contain a small number of *S. aureus*.

Inoculation: Inoculate each of 3 tubes of tryptose soy broth (with 10% sodium chloride and 1% sodium pyruvate) with 1ml of food homogenate. Carry out the same operation from the first and subsequent dilutions using a fresh sterile pipette each time. Maximum dilution of sample must be high enough to yield negative end point. Incubate at 37°C for 48 h.

Surface Streaking: Vortex the tubes 48 h culture in tryptose soy broth and then using 3 mm loop transfer one loopful from each growth positive tube to dried

BP medium plates. Streak to obtain isolated colonies. Incubate at 37°C for 48 hours.

Interpretation: Colonies of *S. aureus* are typically grey black to jet black, circular, smooth, convex, moist, and 2-3 mm diameter on uncrowded BP medium plates. Frequently there is a light colored (off-white) margin, surrounded by opaque zone (precipitate) and frequently with outer opaque zone (precipitate) and frequently with outer clear zone; colonies have buttery to gummy consistency when touched with the inoculating needle.

Confirmation techniques

Using a sterile needle, transfer (noting the dilution) at least one suspected colony from each plate to tubes containing 5 ml BHI and to PCA slants. Incubate BHI cultures and slants at 37°C for 18-24 h. Perform coagulase test on the BHI cultures. Retain slant cultures for repeat tests. Coagulase positive cultures are considered to be *S. aureus*. Record number of positive tubes (and the respective dilutions) of *S. aureus*.

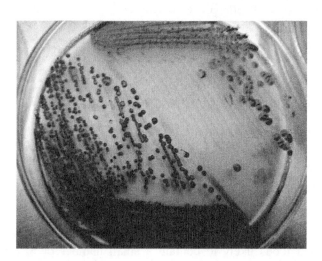

11.11 MOLECULAR SUBTYPING OF *L. monocytogenes* BY PULSED-FIELD GEL ELECTROPHORESIS (PFGE)

PFGE is a powerful technique for molecular subtyping of bacterial pathogens. The technique is used for the separation of large deoxyribonucleic acid (DNA) molecules by applying an electric field that periodically changes direction to a gel matrix. PFGE is essentially the comparison of large genomic DNA fragments after digestion with a restriction enzyme. Since the bacterial chromosome is typically a circular molecule, this digestion yields several linear molecules of DNA.

Preparation of culture

1. Take overnight grown culture of *L. monocytogenes.*
2. Transfer 2 ml of culture to 2 ml of micro-centrifuge tube. (in case of growth on agar media, use sterile polyester-fiber or cotton swab that has been moistened with sterile PBS to remove some of the growth from agar plate; suspend cells in 2 ml of PBS.)
3. Centrifuge the culture at 7500 rpm for 5 min to pellet the cells.
4. Wash the pellet by adding 2 ml of phosphate buffered saline and vortexing.
5. Centrifuge at 7500 for 5 min.
6. Repeat 5 and 6 twice.
7. Suspend in 1-2 ml PBS.
8. Adjust the optical density of culture in the range of 0.79 to 0.81 with the help of PBS.
9. Transfer 240 µl of each bacterial suspension to a 1.5 microcentrifuge tube.
10. Add 60 µl of lysozyme solution (10 mg/ml) and mix by pipetting up and down. Do not vortex.
11. Incubate in a water bath at 37°C for 10 min.

Plug preparation

Preparation: SSP solution

1. Take cell / lysozyme suspensions in 1.5 ml tubes from 37°C water bath.
2. Add 300 ml of the warm (53°C to 56°C) SSP solution to 300 ml cell suspensions; mix by gently pipetting mixture up and down a few times.
3. Immediately, dispense mixture into wells in plug molds. Do not allow bubbles to form. Allow plugs to solidify for 10-15 minutes at room temperature. Alternately, plugs may be solidified by placing in a refrigerator for 5 minutes.

Lysis

1. Prepare cell lysis buffer (50 mM Tris pH 8.0, 50 mM EDTA, pH 8.0, 1% Sarkosyl, 0.15 mg/ml proteinase K)

2. Add 4 ml of cell lysis buffer to each labeled 50 ml polypropylene screw-cap tube.

3. Add plug(s) to tubes containing cell lysis buffer.

4. Incubate plugs in a 54 °C (+/- 1 °C) shaker water bath for 2 h with constant agitation. When making duplicate plugs (two plugs of same isolate), both plugs may be lysed in the same tube.

    Note: All steps up to this point in the protocol should be done in sequence as outlined without delay.

4. Washes

1. Wash plugs 2X with 15 ml of sterile Type 1 water (preheated to 50°C to 54°C) for 10 minutes at 50 °C to 54°C in a water bath with constant shaking.

2. Wash 4X with 15 ml of TE, pH 8.0 (preheated to 50°C to 54°C) for 15 minutes in a water bath with constant shaking.

3. After the last wash is completed, decant TE buffer and add 5 ml of fresh TE, pH 8.0, (room temperature) to each tube.

4. Plugs slices (2-mm) may be restricted immediately with the appropriate enzyme or stored in TE at 4-6°C.

5. Restriction digestion of *L. monocytogenes* DNA in agarose plugs using *Asc*I

1. Dilute 10X Buffer 1:10 in sterile Type 1 water in a labeled tube for the desired number of samples.

    Note: Wear gloves when handling 10X buffer. Keep buffer on ice.

2. Add 150 ml diluted Buffer 4 to each labeled 1.5 ml microcentrifuge tube.

3. Carefully remove plug(s) from tube containing TE with wide end of the spatula. Cut off a 2.0 mm slice and add it to an appropriate labeled 1.5 ml microcentrifuge tube containing diluted Buffer

4.   Be sure plug slice is submerged completely in the buffer. Replace rest of plug in original tube that contains TE Buffer.

5. Place tubes containing plug slices in Buffer 4 in a float and incubate in a 37 °C water bath for 10-15 minutes.

6. While plug slices are incubating in the Buffer 4, prepare the *Asc*1 enzyme mixture.

6. Casting agarose gel

    Preparation: 1% PFGE grade agarose (55-60°C).

7. Assemble gel casting tray, adjust the comb and pour respected amount of molted PFGE grade agarose. Let the agarose solidify (approx. 30 min).

8. Remove plug slices from tubes; put comb on bench top and load plug slices as close to the bottom edge of the teeth as possible

9. Remove excess buffer with tissue. Allow plug slices to air dry on the comb for 5-10 minutes or seal them to the comb with 1% agarose (55-60°C).

10. Carefully pour the agarose (cooled to 55-60°C) into the gel form.

11. Put black gel frame in electrophoresis chamber.

12. Add 2-2.2 L freshly prepared 0.5X TBE. Close cover of unit.

7.   Electrophoresis

1. Set the cooling module at 14°C, power supply, and pump (setting of 70°C for a flow of 1 liter/minute).

2. Remove comb after gel solidifies for 30-45 minutes.

3. Fill in wells of gel with melted and cooled (55- 60°C) 1% SKG Agarose (optional). Unscrew and remove end gates from gel form; remove excess agarose from sides and bottom of casting platform with a tissue. Keep gel on casting platform and carefully place gel inside black gel frame in electrophoresis chamber. Close cover of chamber.

8.   ELECTROPHORESIS CONDITIONS

Use the following electrophoresis conditions for *Asc*1 or *Apa*1 digested *L. monocytogenes* DNA plugs slices when using the Chef Mapper:

Running buffer 0.5X TBE; temperature = 14 °C

Gel 1.0% PFGE grade agarose prepared in 0.5X TBE

Change initial switch time to = 4.0 seconds

Change Final switch time to = 40.0 seconds

V 6 V/cm

Run 22 hrs

Press "Start Run"

Staining and documentation of PAGE agarose gel

1. When electrophoresis run is over, turn off equipment; remove and stain gel with ethidium bromide by diluting 40 µl of ethidium bromide stock solution (10 mg/ml) with 400 ml of reagent grade water (this volume is for a staining box that is approximately 14 cm x 24 cm; a larger container may require a larger amount of staining solution). Stain gel for 20 - 30 min in covered container.

Note: Ethidium bromide is toxic and a mutagen; the solution can be kept in dark bottle and reused 5 - 6 times before discarding according to your institution's guidelines for hazardous waste or use the destaining bags recommended for disposal of ethidium bromide.

2. Destain gel in approximately 500 ml reagent grade water for 30-60 min. Capture image on gel documentation system. If background interferes with resolution, destain for an additional 30-60 min.

    For different bacterial pathogens, the restriction enzymes, plug preparation and running conditions differ. The details of the PFGE conditions for different pathogens are given in the following Table.

11.12 INTRODUCTION TO MATRIX ASSISTED LASER DESORPTION IONIZATION TIME OF FLIGHT MASS SPECTROMETRY

MALDI-TOF MS is an innovative tool that is easy to use, rapid, accurate, and cost-effective, and that has revolutionized bacterial identification in clinical microbiology laboratories. MALDI-TOF MS allows the identification of microorganisms at the species level by analysis of total protein. This technique is based on the generation of mass spectra from whole cells and their comparison to reference spectra. Once the samples are ready, species determination takes only a few minutes. Hence, MALDI-TOF MS has been integrated with in the last decade into the routine work flow of medical microbiology laboratories for microbial identification, replacing the traditional biochemical or molecular techniques.

The Principle

This technique quickly identifies bacterial species by determining the molecular masses of proteins, many of which are ribosomal proteins. The analysis is performed on whole cells or on crude bacterial extracts. MALDI-TOF MS requires a stage of sample crystallization in a matrix on an inert support. The irradiation of the crystalline mixture with a brief laser pulse desorbs and ionizes characteristic ions (MALDI). The ionized molecules are then accelerated in an electric field and separated, as a function of molecular weight, in a vacuum tube. The time-of-flight (TOF) of each ion is measured and corresponds to the time taken to reach the detector. The ions reach the detector, more or less rapidly, according to their m/z ratio. The electrical signal is processed by computer and converted into a mass spectrum. The mass range acquisition is mainly m/z 2000 to 20000.

PFGE conditions used for different bacterial pathogens

Organism	Enzyme (s)used	Enzyme conc.	Running conditions CHEF DRII	CHEF DR III	Source
Listeria monocytogenes	Asc I Apa I	25 U/sample 160 U/sample	Initial switch time: 4.0 s Final switch time: 40.0 s Voltage: 6 V Included angle: 120° Run time: 18-19 hours	Initial A time: 4.0 s Final A time: 40.0 s Start ratio: 1.0 (if applicable) Voltage: 200 V Run time: 19-20 hours	CDC-Pulsenet
Clostridium botulinum	Xba I Xha I Sma I	50 U/sample 100 U/sample 50 U/sample	Initial switch time: 0.5 s Final switch time: 40 s Voltage: 6 V Included angle: 120° Run time: 18-19 hours	Initial switch time: 0.5 s Final switch time: 40 s Voltage: 6 V Included angle: 120° Run time: 18-19 hours	CDC-Pulsenet
Escherichia coli O157:H7	Xba I Bln I	50 U/sample 30 U/sample	Initial A time: 2.2 s Final A time: 54.2 s Start ratio: 1.0 (if applicable) Voltage: 200 V Run time: 19-20 hours	Initial switch time: 2.2 s Final switch time: 54.2 s Voltage: 6 V Included angle: 120° Run time: 18-19 hours	CDC-Pulsenet
Escherichia coli non-O157 (STEC)	Xba I Bln I	50 U/sample 30 U/sample	Initial A time: 6.76 s Final A time: 35.38 s Start ratio: 1.0 (if applicable) Voltage: 200 V Run time: 19-20 hours	Initial switch time: 6.76 s Final switch time: 35.38 s Voltage: 6 V Included angle: 120° Run time: 18-19 hours	CDC-Pulsenet
Salmonella serotypes	Xba I Bln I	50 U/sample 30 U/sample	Initial A time: 2.2s Final A time: 63.8 s Start ratio: 1.0 (if applicable) Voltage: 200 V Run time: 19-20 hours	Initial switch time: 2.2 s Final switch time: 63.8 s Voltage: 6 V Included angle: 120° Run time: 18-19 hours	CDC-Pulsenet

[Table Contd.

Contd. Table]

Organism	Enzyme (s) used	Enzyme conc.	Running conditions		Source
			CHEF DRII	CHEF DR III	
Shigella sonnei and Shigella flexneri	Not I	50 U/sample	Initial A time: 2.2 s Final A time: 54.2 s Start ratio: 1.0 (if applicable) Voltage: 200 V Run time: 19-20 hours	Initial switch time: 2.2 s Final switch time: 54.2 s Voltage: 6 V Included angle: 120° Run time: 18-19 hours	CDC-Pulsenet
	Xba I	50 U/sample			
	Spe I	30 U/sample		Initial switch time: 6.8 s Final switch time: 35.4 s Voltage: 6 V Included Angle: 120° Run time: 18-19 hours	CDC-Pulsenet
Campylobacter jejuni	Sma I	40 U/sample	Initial A time: 6.8 s Final A time: 35.4 s Start ratio: 1.0 (if applicable) Voltage: 200 V Run time: 19-20 hours		
	Kpn I	40 U/sample			
Cronobacter spp.	Xba I	50 U/sample	Initial A time: 1.8 s Final A time: 25 s Start ratio: 1.0 (if applicable) Voltage: 200 V Run time: 19-20 hours	Initial switch time: 1.8 s Final switch time: 25 s Voltage: 6 V Included Angle: 120° Run time: 18-19 hours	CDC-Pulsenet
Vibrio cholerae and Vibrio parahaemolyticus	Sfi I	40 U/sample	Initial A time: 10s Final A time: 35 s Start ratio: 1.0 (if applicable) Voltage: 200 V Run time: 19-20 hours	Initial switch time: 10 s Final switch time: 35 s Voltage: 6 V Included angle: 120° Run time: 18-19 hours	CDC-Pulsenet
	Not I				
Yersinia pestis	Xba I	50 U/sample	Initial A time: 1.79 s Final A time: 18.66 s Start ratio: 1.0 Voltage: 200 V Run time: 20-22 h	Initial A time: 1.79 s Final A time: 18.66 s Start ratio: 1.0 Voltage: 200 V Run time: 18-20 h	Asc I Pulsenet
		40 U/sample			

[Table Contd.

Contd. Table]

Organism	Enzyme (s)used	Enzyme conc.	Running conditions		Source
			CHEF DRII	CHEF DR III	
Staphylococcus aureus	Smal	30 U/sample	Volts = 200 (6 v/cm) Temp = 14 °C Initial switch = 5 seconds Final switch = 40 seconds Time = 21 hours for SeaKem Gold agarose	Volts = 200 (6 v/cm) Temp = 14°C Initial switch = 5 seconds Final switch = 40 seconds Time = 21 hours for SeaKem Gold agarose	CDC-Pulsenet
Brucella spp.	Xbal	30 U/sample	Run time 21 h Pulse time of 0.5-8.0 s Angle of 120 °C, Gradient of 6.0 V/cm, Temperature of 14 °C	Run time 21 h Pulse time of 0.5-8.0 s Angle of 120 °C, Gradient of 6.0 V/cm, Temperature of 14 °C	Brower et al., 2007

❑ Mass spectrometry is an analytical technique for determining the elemental composition of a sample.

❑ Principle: ionizing chemical compounds to generate charged molecules and to measure their mass-to-charge ratio.

❑ Rapid identification of bacteria/yeast/fungi, directly from colonies, by MALDI-TOF mass spectrometry (Matrix assisted laser desorption/ionisation - Time of Flight).

❑ In bacterial identification, MALDI-TOF examines the patterns of proteins detected directly from intact bacteria.

Procedure

1. The target slide is prepared and introduced into a high-vacuum environment. A precise laser beam ionizes the sample and laser light causes the sample and matrix to volatilize. The time of flight of the proteins is recorded using a formula from the time recorded.

2. Proteins are detected with a sensor to create a spectrum that represents the protein makeup of each sample.

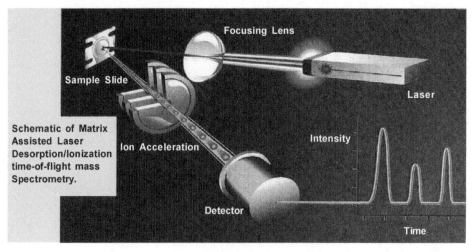

✓ The database is for worldwide pharma/environment/food sectors.

✓ The database consists of some of the most specific species

✓ Molds are also now possible.

✓ Rapid, green, robust, accurate and still evolving

✓ Automated, reliable and single colony requirement

✓ Low exposure risk – sample inactivation

✓ Soft ionization technique- enables molecules to remain intact during the procedure.

✓ Minimal cost of consumables per specimen compared with conventional methods. Do not have to perform time-consuming, costly and labor intensive phenotypic and molecular typing tests like staining, microscopy, motility, catalase / oxidase tests, 16sr RNA sequencing, PFGE, AFLP, MLST etc.

✓ Increasingly used in clinical laboratories worldwide

✓ Detection of antibiotic resistance

✓ Fast and definitive cause of infection gives patient immediate clinical help.

✓ Inter-laboratory reproducibility

✓ Adaptive open system and expandable by user

✓ Broad applicability- gives most bacterial (aerobes and anaerobes) and fungal identification of isolates

✓ Determination using whole cell, cell lysates or crude extracts

✓ Highly abundant ribosomal proteins can be measured.

Requirements

1. Matrix: CHCA and formic acid
2. MS worksheet
3. MS slide
4. Freshly grown standard culture and samples on plates
5. 1 µl loop and a wooden toothpick
6. Disposable gloves, mask, lab coat
7. Autoclaved tips, pipettes and 70% ethanol

Protocol

1. Take matrix and formic acid out of the fridge and warm up in room temperature on bench.
2. Label the MS worksheet with the date and slide number.
3. Apply a fresh isolate (24 hrs) of standard culture control in the central circle of the slide using 1µl loop.
4. Add 1 µl of matrix. (Do not invert the matrix vial. Do not touch the crystals on the bottom of the vial with pipette tip).

5. Ensure plates to be tested match the accompanying label.

6. Place one specimen barcode label on the worksheet.

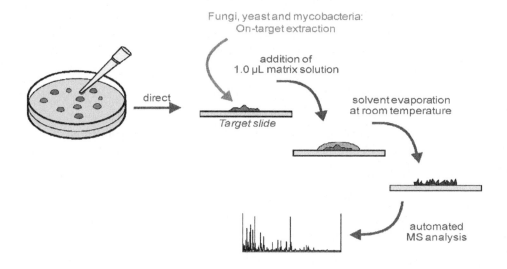

7. Apply organism in each circle of the slide by using 1 μl of loop or wooden toothpick.

8. Add 1ul of Matrix.

   (For Yeast, add 0.5 μl of formic acid first, wait until it crystallizes, then add 1 μl matrix).

9. Acquire spectra.

11.13 ELECTRO IMMUNO TRANSFER BLOT (EITB) OR WESTERN BLOT

The western blot (sometimes called the protein immunoblot) is a widely used analytical technique in molecular biology, immunogenetics and also in meat science. Specific or polyclonal antibodies are created that react with a specific target protein. The sample material undergoes protein denaturation, followed by gel electrophoresis. Later, the electrophoresis membrane is washed in a solution containing the specific antibody. The excess antibody is then washed off, and a secondary antibody that reacts with the first antibody is added. Through various methods such as staining, immunofluorescence, and radioactivity, the secondary antibody can then allow visualization of the protein

Initially, sodium dodecyl sulphate polyacrylamide gel electrophoresis (SDS-PAGE) is performed (described earlier)

MALDI-TOF MS

A rapid method for identification of bacterial and fungal strains

Five minute protocol

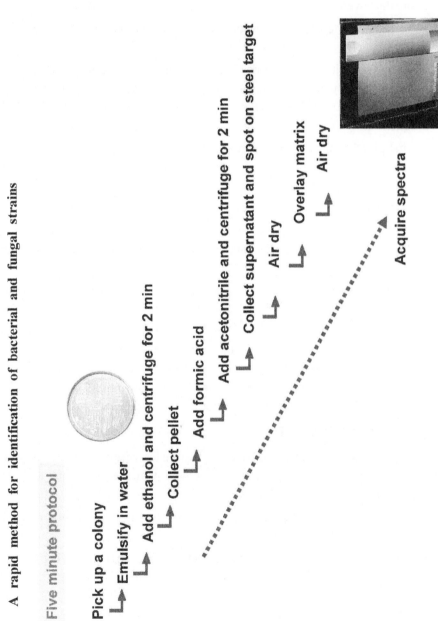

Pick up a colony

↳ Emulsify in water

↳ Add ethanol and centrifuge for 2 min

↳ Collect pellet

↳ Add formic acid

↳ Add acetonitrile and centrifuge for 2 min

↳ Collect supernatant and spot on steel target

↳ Air dry

↳ Overlay matrix

↳ Air dry

Acquire spectra

1. SDS-PAGE is performed using the different concentrations of antigens viz. 30 μg, 35 μg and 40 μg in the wells of gel slab. Pre-stained protein molecular marker (broad range) with protein bands of 180, 116, 90, 58, 48.5, 36.5 and 26.6 kDa molecular weight is used.

2. After the completion of electrophoresis, the gels are washed thoroughly with distilled water and carefully remove from the glass cassette.

3. To transfer the protein on a nitrocellulose membrane, the Blot Dry blotting system gel transfer device is used.

4. To use the Blot Dry Blotting system for protein transfer, assemble the Blot gel transfer stack containing the nitrocellulose transfer membrane with pre electrophoresed gel. Any trapped air can be removed by the De-bubbling roller.

5. First place the anode stack (bottom), contains a copper electrode, nitrocellulose membrane and bottom transfer gel layer packaged in a transparent plastic tray.

6. Then place the pre electrophoresed gel and put the wet filter paper on the gel.

7. Place cathode stack (top) contains a copper electrode and top transfer gel layer packaged in a red plastic tray.

8. Lastly, the disposable sponge is placed on the inner side of the blot lid within the small protrusions on the lid. The sponge absorbs any excess liquid on the stacks during blotting and generates an even pressure on stack assembly.

9. After completion of the program, remove the gel. The prestained marker can be seen on the nitrocellulose membrane which indicates the complete transfer of bands from gel to a nitrocellulose membrane.

Western Blotting Assay

1. After the completion of immunoelectron transfer, carefully remove the sponge, the cathode stack, filter paper and the gels with use of forceps and discard properly.

2. Cut the nitrocellulose membrane carefully into stripes according to the lanes in which the protein was loaded.

3. The enzyme-linked immunosorbent assay will be carried out. The stripes are submerged in 5% skimmed milk solution in PBS (blocking buffer), approximately 10 ml into sterile petri dish. The petri dishes kept for incubation @ 37°C for 2 hours in a shaker incubator with constant shaking.

4. Then washed stripes thrice with washing buffer (PBS-T) and incubated with serum (hyper-immune rabbit serum / pig serum / human serum) diluted in PBS in ranges from 1:100, 1:75 and 1:50 for 2 hours @ 37°C with intermittent shaking.

5. After treatment with the primary antibody, the stripes again washed with PBS-T and incubated with anti-species IgG (whole molecule) peroxidase conjugate (Sigma Aldrich, USA) diluted in PBS (pH 7.2-7.4) in ranges of 1:20000, 1:10000 and 1:5000, for 1 hour @ 37°C in shaker incubator.

6. Wash stripes thrice with washing buffer (PBS-T) and put in the substrate solution for 5-7 minutes. The substrate solution comprises of 3', 3'-diaminobenzidine- tetrahydrochloride diluted in PBS (pH 7.2-7.3) and with (5-10µl) of 30% H_2O_2 .

7. Stop the reaction by washing stripes with distilled water.

8. A positive reaction is confirmed by the appearance of definite dark brown colored bands on the stripes as the immunoreactive bands.

9. The relative molecular weight of the bands is determined by comparing with the standard molecular weights on the pre-stained protein marker. The relative comparison is done using the Quantity-one software in gel documentation system.

Reference

Bureau of Indian Standards (BIS) have also suggested the method for plate count of bacteria in food stuff as per I.S. 5402: 1969.

CHAPTER - 12

■ ■ ■

PHYSICO-CHEMICAL TECHNIQUES FOR SPECIES IDENTIFICATION OF MEAT

The characteristics of meat and fat of food animals depend mostly on their colour, texture, appearance, consistency and distribution of fat. The differentiation of the muscle and fat of animals is of importance in connection with the possible substitution of inferior and at times repugnant meat for that of good quality. The substitutions that may be practiced are, that of horseflesh for beef, mutton for chevon, mutton for venison, beef for mutton and occasionally the flesh of the cat for that of hare or rabbit. It is not difficult to differentiate the flesh and fat of these animals in the carcass form or in joints by means of anatomical conformation. But, their recognition in minced meat or in product form depends on chemical and biological tests.

12.1 PHYSICAL METHOD

Physical methods like anatomical differences of each species of the carcass and appearance of muscle and fat colour, odour, texture and taste have provided a general difference between species in earlier days for food analysis. So, this can be attempted, provided the meats are in the form of joints and in carcass form.

Beef

It is light red in young and dark and coarser in older animals due to testosterone secretion. Intramuscular fat (marbling) is well marked especially in well fed animals. In old bulls the dried surface of meat appears darker. Fat is yellowish in colour due to carotene pigment and firm in consistency. In older animals the fat tends to be more yellowish and loose in consistency.

Veal

In very young calves it is pale in colour, watery in consistency and fat is white in colour and jelly like.

Carabeef (buff)

Generally buffalo meat is darker (more reddish brown) and the fibres are coarser and looser in structure than beef. The odour of the buffalo meat and fat resembles that of musk and if boiled in strong acidified (H_2SO_4) water, it develops a disagreeable odour similar to that of cattle manure. The fat of buffalo is strikingly white and dry and less sticky than in cattle. There is no marbling seen in buffalo meat.

Mutton

Mutton is light red in colour with fine firm muscle fibres with little or no marbling. Fat is deposited in the fat depots and is white in colour, odourless, firm. Feathering, a characteristic phenomenon of fat distribution in intercostal muscles is a feature of mutton carcass.

Chevon

Chevon is almost like mutton but pale in colour with no marbling but depot fat is marked white colour and has goaty odour. Kidney fat is abundant. Goat meat is more lean, dry and firm and easy to cook.

Pork

It is light to dark in colour, varying from whitish grey to red, less firm in consistency and has a strong boar odour. Subcutaneous fat deposition is predominant which is white in colour and greasy in texture. Pork becomes nearly white on cooking while other meats become darker.

Chicken

Meat of a broiler chicken is pale brown in colour; fat is greasy and shows marbling. The subcutaneous fat is abundant in broiler than desi type. Meat gives a pleasant odour.

Turkey

Meat is more pinkish than broiler poultry and has a pleasant meaty odour. Dressed turkey has more breast and less thigh meat than other table birds.

Rabbit

Farm-raised rabbit is lean, slightly sweet meat with a closely textured flesh that has virtually no fat and is very high in protein. Rabbit meat does not have a very strong flavour. It is comparable to, but not identical with chicken. Tenderness varies with muscle age, and depends on changes in the proportion and type of connective tissue supporting the muscle fibres. The younger rabbit meat is tenderer. The typical strong flavour tends to develop with age.

Horse

Horseflesh is dark red even bluish in colour and has more fascia (due to more exercise) with sweet or somewhat repulsive odour, which develops rusting on keeping. There is no marbling and fat is yellow and greasy.

Deer

In deer, the subcutaneous layer of fat is not as well developed as in sheep. The meat possesses the odour of venison, which is easily distinguishable from the odour of sheep meat.

Dog

The colour of dog meat is very dark than pork and is easily made out in cooked meat. The muscles of the dog are smeary and the fat is oilier than hog fat. The odour of the dog meat is repulsive

12.2 Chemical Tests

The Chemical test consist of the determination of

a. The content of glycogen in flesh
b. Myoglobin content
c. The percentage of linoleic acid in fat,
d. The amount of iodine absorbed by unsaturated fatty acids in fat (iodine value)
e. The refractive index of fat.

Table 12.1: Differentiation of carcasses of horse and cattle based on anatomical differences

Horse	Cattle
Unusual length of sides, together with the great muscular development	Beef cattle are generally short and stocky
The thoracic cavity is longer and has 18 pairs of ribs	13 pairs of ribs
The ribs are narrower but more markedly curved	Ribs are flat
The superior spinus processes of the first six dorsal vertebrae are more developed and less inclined posterior.	It extends and articulates with the carpus
In the forequarter, the ulna extends only half the length of the radius	Femur possesses no third trochanter; the fibula is only a small pointed projection.
In the hind quarter, the fibula extends two thirds the length of the tibia	They do not articulate
The last three lumber transverse processes articulate with each other, the sixth articulating in the similar manner with the sacrum	The kidney fat is always firmer, white and more abundant than in horse.
Carcass shows considerable development of soft yellow fat beneath the peritoneum, especially in mares, but in stallion the fat is generally of a lighter colour and almost white.	No bluish tinge of muscle
Flesh is dark bluish red. Meat has pronounced sweet taste and well-defined muscle fibres.	

Table 12.2: Differentiation of carcasses of sheep and goat based on anatomical differences

Feature	Sheep	Goat
Back and withers	Round and well fleshed	Sharp, little flesh
Thorax	Barrel shaped	Flattened laterally
Tail	Fairly broad	Thin
Radius	1.25 times length of metacarpus	Twice as long as metacarpus
Scapula	Short and broad. Superior spine, bent back and thickened	Possesses distinct neck. Spine straight and narrow
Sacrum	Lateral borders thickened in form of rolls	Lateral borders thin and sharp
Flesh	Pale red and fine in texture	Dark red and coarse with goaty odour. Sticky subcutaneous tissue, which may have adherent goat hairs.

12.2.1 Glycogen Content of Meat

The horseflesh is richer in glycogen than the flesh of other food animals irrespective of age.

Horse: 0.5 to 1.0 percent, beef: up to 0.5 percent, pork and mutton: 0.2 percent.

Caution

(i) The flesh should be tested for the content of glycogen soon after the slaughter as it disappears from the flesh quickly.

(ii) Liver of all food animals especially pig liver contains more glycogen, when they are used in sausage making, it gives a high percentage. So care must be taken in interpretation of results.

12.2.2 Myoglobin Content

Myoglobin is a pigment responsible for giving red colour to the meat.

Horse	0.71 percent
Cattle	0.60 percent
Pig	0.43 percent
Sheep	0.35 percent
Pigeon	0.22 percent
Rabbit	0.02 percent

Source: - Lawrie, R.A. (1952), Nature, 170, 4316.

12.2.3 Linoleic Acid Content

Linoleic acid content of horse fat is 1to2 percent , which is much higher than other animals (not higher than 0.1 percent.)

12.2.4 Iodine Value

This test is based on the amount of iodine absorbed by the unsaturated fatty acids present in the fat and varies in different animals. The iodine value of animal fats are- (i) Horse fat: 71 to 86, (ii) Beef fat: 38 to 46, (iii) Sheep and Goat fat: 35 to 46 (iv) Pig fat: 50 to 70.

The iodine value of ox and sheep are to some extent found to be closely identical. Hence, it is not possible to confirm the type of meat legally based on iodine number.

12.2.5 Refractive Index

In this method the fat is liquefied by heating and its refractive index is estimated. All liquids including oils possess a specific refractive index and it is estimated by refractometer. (i) Horse fat: 53.5 (ii) Ox fat: not above 40 (iii) Pig fat: not above 51.9

12.3 BIOLOGICAL TESTS

Generally, these methods are considered to be a foolproof methods in judicial enquiry and particularly, applicable to minced meat or processed meat. Among the various tests, the precipitation test is the most valuable one. It is based on the development of certain antibodies in the blood serum from another animal and identification of species-specific protein fraction.

12.3.1 Electrophoresis Techniques

Electrophoresis separation of proteins on the basis of their mobility (molecular weight) in a supporting gel of agarose at a constant pH and electrical field is termed as agar gel electrophoresis. These separated meat protein fractions aid in identifying origin of meat.

Gel electrophoresis

Gel electrophoresis is an adaptation of electrophoresis. Instead of filter paper, a gel medium is used containing a hydrophobic medium as internal phase and an electrolyte buffer as the external or solvent phase. In gel electrophoresis, there are two factors, which tend to separate protein or nucleic acid fractions. They are: (i) The application of an electric current which produces separation on the basis of charge density (rate of migration in the electric field) and (ii) Molecular sieving (separation based on pore size of the material though which the components of a mixture migrate). Thus an increase in concentration of the solid phase in the gel will result in a decrease in the pore size and greater retention of large molecules.

Three types of gels are used for protein electrophoresis, (i) Agar gel is most suitable for electrophoresis of protein containing immune factors, as their transparency permits recognition of antigen-antibody complexes; (ii) Starch gel, acts more on the basis of molecular sieving whereas, (iii) Acrylamide gel is most useful when both molecular sieving and rate of migration in the electric field are desired. After separation the gels are stained to locate the bands and the excess dye removed by destaining of gels.

Feasibility of Technique

While determining the protein pattern specific to the species, one has to consider whether these differences are species specific or an expression of different conditions like age of animal, nutritional status and influence of stress. Post mortem changes like rigor mortis also has effect on state of protein and extent of denaturation. Conditions like pale, soft and exudative (PSE) or dark, firm and dry (DFD) also alter the protein solubility. These changes would influence the electrophoretic pattern in each animal and has to be considered while applying the test. Hence the test has limited value in field conditions.

12.3.2 Iso-electric Focussing (IEF)

This technique is better in higher resolution and reproducibility over the electrophoretic method. The IEF is an electrophoretic technique, which utilises the charge on the surface of the protein molecule to drive it through a gradient gel. The pH gradient is set up by polymeric compound ampholytes. The proteins applied on to the gel reach a point when the surface charges become neutral at their isoelectric point. Proteins are fixed as bands. Bands vary with species. The proteins are detected with staining and densitometric scanning by counting the number of bands, their distribution pattern, determining PI (iso electric point) values and marking the zones. Thus, it is possible to have a specific focusing pattern for each animal species.

12.3.3 Immunological Techniques

In the immunological methods specific proteins can be characterised without resolution using specific immuno reagents.

In this method, the blood serum from the specific immunised rabbits is mixed in a test tube with a filtered extract from the suspected meat. A turbidity of the solution first occurs if the specified meat is present and this is then followed by definite precipitation. The precipitation test is a protein reaction of value for detection of flesh, organ fat or intestines. It is of the greatest value with raw meat or if meat is very well boiled or fried.

Based on a simple double diffusion method (Ouchterlony, 1948) specific anti-species serum (antibodies) and unidentified meat extract (antigen) are allowed to diffuse towards one another in agar gel slab. If the antigen and antibody are homologous, a precipitin band is formed along the line where the two meet.

Prominent techniques are:

- **Precipitating tests in the tubes**

  This test is also called single gel diffusion or immunodiffusion test where the antigen is layered over the antiserum in capillary tube. The reaction occurs at the interface of the layers in the form of immuno precipitate.

- **Double immunodiffusion test / Agar gel precipitation test (DID / AGPT)**

  This technique is more sensitive than single diffusion technique and has number of advantages. Hence, it is widely used and accepted for identification of species origin of meat samples as well as for the detection of adulteration of one species of meat with other species in the field.

  In this method antigen and antibody both are allowed to diffuse towards each other in an inert medium e.g. agar gel.

  Presence of reaction of identity (complete fusion / merger or precipitation are formed due to reaction between test antigen and antiserum with the one due to reaction between positive control and the antisera) is the criterion to identify the test antigen as identical to the known antigen. The test antigen that forms precipitation arcs but without the reaction of identity is not considered as identical to the known standard antigen. The gels can be stained by amido black.

Feasibility of the Test

The technique is simple, rapid, economic and easy to perform in the field.

The results can be provided in 24 hours and adulteration to a level of 5-6 percent can be detected.

The resultant lines of precipitation do not diffuse after formation and can be preserved for future reference by staining.

The test can be performed on putrid samples without having adverse effect on results.

The test can be used under field conditions for raw, chilled or frozen meat.

The test can also detect the additions in case of boiled and autoclaved meats and organs using their BE (Boiled and ethanol precipitated) forms within 24 hours.

Limitations

Accuracy of the test depends mainly on the potency (high titer) and species specificity of diagnostic antibody.

Cross-reacting antiserum may pose problem in species specific detection.

12.3.4 ELISA test

It is a powerful qualitative as well as quantitative immunological test. Two versions of this test as solid phase indirect and capture ELISA are used. The indirect ELISA uses meat extracts to coat the solid phase (polyvinyl chloride). The species-specific antibody added to the wells recognises species antigen adsorbed to the PVC and the enzyme labeled anti-species IgG detects species-specific antibody. The capture assay employs a antibody adsorbed to the PVC wells to bind meat antigen and enzyme labeled species specific antibody to detect the presence of bound antigen,

ELISA can detect the adulterants to a minimum level of 1 percent in case of all closely related animal species by using simple water extracts as the antigens.

The PCR technique in meat species identification is more specific and sensitive.

■ ■ ■

MOLECULAR TECHNIQUES FOR SPECIES IDENTIFICAITON OF MEAT

In the last two decades, several molecular based techniques have been employed by researchers for various applications in meat science. Molecular biology techniques depend on analysis of DNA to solve the specific problem. One important preliminary step in meat species identification using molecular techniques is extraction of nucleic acids, especially DNA, from the collected samples. Every living cell has nucleic acids as an essential entity of life. Nucleic acids were discovered by Friedrich Mieshner in 1871, since they were found in cell's nucleus, the name 'nucleic acid' was given. Apart from nucleus of the cell, nucleic acids are also found in mitochondria (animal cell) and chloroplasts (plant cell). Availability of quality DNA free from contaminants is a basic prerequisite for undertaking molecular techniques.

13.1 DNA EXTRACTION FROM MEAT SAMPLES

13.1.1 Extraction of DNA from meat by Phenol: Chloroform: Isoamyl alcohol method

Principle

Phenol: Chloroform: Isoamyl alcohol (PCI) method is based on the principle of liquid-liquid partition (LLP). In LLP method, the DNA is partitioned between two immiscible solvents *i.e.* an aqueous (upper portion containing DNA) and organic (lower portion with impurities) phases. The DNA is water soluble and hence it gets partitioned into upper (aqueous) phase; while other impurities such as tissue matrix (proteins) precipitate and get partitioned into the lower organic phase (discarded). Extracted DNA is precipitated from the aqueous phase using absolute alcohol (ethanol) or isopropyl alcohol and precipitated DNA is pelleted by centrifugation (alternatively, DNA threads can be spooled using glass rod). The DNA pellet is dried and dissolved in tris-ethylene diamine tetra acetate (TE buffer) or nuclease free water (NFW) and stored in deep freezer (-20°C) until further use. The phenol-chloroform extraction is a simple and economical method; hence, it is most commonly used in molecular biology laboratories.

Reagents required

- Lysis buffer
- Proteinase K enzyme
- Tris saturated Phenol
- Chloroform
- Isoamyalcohol
- Ammonium acetate solution (1M)
- Ethanol
- Tris - EDTA buffer

Procedure

- Mix about 75 mg crushed/minced fresh meat with ten volumes of lysis buffer (fresh meat), pH 8.0. Incubate at 37°C for 1 hr.
- Add Proteinase K enzyme (0.1 mg/ml) and incubate at 50°C for 3 hrs.
- Mix incubated solution with equal volume of Phenol/ Chloroform / Isoamyalcohol (25:24:1).
- Centrifuge at 3000 rpm for 2-3 min.
- Transfer aqueous phase and mix with equal volume of chloroform.
- Centrifuge at 3000 rpm for 2-3 min.
- Transfer aqueous phase and mix with equal volume of chloroform.
- Centrifuge at 3000 rpm for 2-3 min.
- Mix aqueous phase with 0.2 volumes of 1M ammonium acetate solution and two volumes of ethanol. Shake thoroughly.
- Centrifuge the solution at 10000 rpm for 5 min.
- Discard supernatant to get the DNA pellet.
- Wash DNA with 70% alcohol and centrifuge at 10000 rpm for 5 min.
- Mix DNA pellet with IX TE (Tris- EDTA) buffer.

- Keep DNA solution in water bath at 60°C for 2 hr to inhibit DNase activity and to enable proper dissolving of pellet in the TE buffer.
- Store at −20°C for further use.
- For processed meats, procedure described is same except that lysis buffer (processed meats) is to be used in first step.

Composition of reagensts required for DNA extraction

Lysis buffer (cooked meat)

- Tris-Cl (pH 8.0)	10 mM
- NaCl	100 mM
- SDS (w/v)	1%
- Di-thiothreitol (DTT)	10mM

Lysis buffer (fresh meat)

- Tris-Cl (pH 8.0)	10 mM
- EDTA (pH 8.0)	0.1 M
- SDS (w/v)	0.5%
- RNase (DNase free, pancreatic)	20μg/ml

Proteinase K

- Dissolve 10 mg of Proteinase K in 1 ml of water and store in aliquots at -20°C

  i.e., 10 mg/ml à 10,000 μg/1,000 μl à 10 μg/μl Indicating,

  1 μl = 10 μg

  2 μl = 20 μg

  20 μl = 200μg
- Add Proteinase K @ 200 μg/ml

Ammonium acetate (10 mM)

- Ammonium acetate	77.08 mg
- Distilled water (up to)	100 ml

Ethylene diamine tetracetate (EDTA), 0.5 M, pH 8.0

- EDTA	18.612 g
- Distilled water	80 ml

- Stir vigorously on a magnetic stirrer to mix.
- Adjust the pH to 8.0 and make the volume to 100 ml.
- Autoclave and store

Tris-EDTA (TE) buffer 1X, pH 8.0

- Tris HCl   10mM
- EDTA   1 mM
- Distilled water up to 100 ml and adjust pH 8.0

13.1.2 Extraction of DNA from meat by Alkaline lysis method

This method will help in rapid extraction of DNA using simple equipments and reagents thereby it reduces cost of extraction and brings down the time required for analysis. In tests involving export consignments, increased period of holding of consignments in cold storage will have economic implications on the consignee. Hence, it is essential to provide the results of the assay at the earliest to reduce the period of holding. Alkaline lysis method of DNA extraction is rapid as the entire process of DNA extraction takes less than 30 min, economical as it involves only less expensive chemicals and simple as it does not require any sophisticated equipments. It can be undertaken by even semi skilled person as the method is less cumbersome. Further, minimal pipetting steps are involved thus reducing the chances of cross contamination of samples and cost on plastic wares. However, yield of the DNA will be low but will be sufficient for undertaking PCR.

Principle

Alkaline lysis method works on the principle that alkaline pH and high temperature lyses cells & degrades protein leaving DNA in the solution.

Reagents

- 0.2 N NaOH
- 0.04 M Tris HCl

Procedure

1) Triturate 500 mg of meat in pestle & mortar with eight volumes (4 ml) of 0.2 N NaOH.
2) Take 5 μl of the extract and mix with eight volumes (40 μl) of 0.2N NaOH in a sterile micro centrifuge tube.
3) Heat the mix in water bath at 75°C for 20 min

4) Add eight volumes (360 µl) of 0.04 M Tris HCl (pH 7.75) for neutralization of pH.

5) 1 µl of final mix containing about 100 ng DNA can be used for PCR (Grish *et al.*, 2013)

13.1.3 Checking quality, purity and concentration of DNA

Quality of DNA: Quality of DNA can be evaluated by electrophoresing in 0.8 % agarose gel. Intact DNA without shearing will show single band upon electrophoresis while sheared DNA will form a smear.

Purity of DNA: The purity of DNA is checked using UV Spectrophotmetry. Optical density of the DNA samples is measured at 260 and 280nm. OD ratio (260:280) between 1.7 and 1.9 is considered good and can be used for PCR reaction. Those samples showing value, beyond this range need to be reprocessed.

Concentration of DNA: Concentration of DNA can be estimated using following formula:

$$\text{DNA concentration } (\mu g/\mu l) = \frac{\text{OD260} \times \text{Dilution factor} \times 50}{1000}$$

(1 OD value at 260 nm is equivalent to 50 ng dsDNA/µl)

13.2 VISUALIZATION OF DNA BY AGAROSE GEL ELECTROPHORESIS

Agarose gel electrophoresis is the commonly used method for visualization of DNA fragments and PCR amplicons. DNA has to be resolved (*i.e.* separation of different molecular weight fragments) and it is undertaken by agarose gel electrophoresis process. It is required mainly for checking quality of isolated DNA, resolution of PCR products and RFLP fragments etc.

Principle

The DNA molecule has a net negative charge and hence it migrates towards positive electrode in buffer solution. The 'gel' is a matrix that acts as a 'molecular sieve' for the charge dependent migration of DNA. The sample DNA is loaded into wells created in the gel that is immersed in the buffer solution. The DNA molecule gets resolved depending upon its molecular weight (size/ length in base pairs). Under the influence of electric field, rate of migration of DNA molecule is inversely related to the DNA size (*i.e.* lower the molecular weight, faster will be the migration and *vice-versa*). However, such DNA cannot be visualized by an unaided (naked) eye; it requires staining of DNA for visualization. Intercalating dye such as ethidium bromide bind to double stranded DNA and when UV-rays are bombarded over it, the dye intercalated in the DNA emits luminescence and hence presence of DNA becomes evident.

Reagents required

1) TAE buffer (50x) Prepared by mixing Tri base (12.1 g), Acetic acid (2.8 ml), 0.5 MEDTA, pH 8.0 (5 ml) with distilled water to make up to 50 ml

2) Agarose

3) Ethidium bromide dye

4) Loading dye

Procedure

1) **Gel casting:** Prepare agarose mixture of required concentration in buffer (TAE 0.5x). Warming to dissolve agarose completely in buffer and cool to 40-50°C. Add ethidium bromide dye @ 10 mg/ml. Mix thoroughly for uniformity. Pour gel into pre-positioned gel casting tray and allow to solidify. Ensure thickness gel to about 4 mm. Remove the comb and tray accessories from gel casting tray. Placing the gel (along with side solid support) into the buffer tank of the submarine gel electrophoresis unit.

2) **Sample loading:** Wells are located near negative electrode as DNA is negatively charged it migrates towards positive electrode (cathode). Load DNA samples mixed with loading dye (6x gel loading dye) with controls (positive and negative control and marker (ladder) at either ends of gel. Close the buffer tank lid.

3) **Electrophoresis:** Run power pack so as to discharge a constant voltage (50 V) for a specified duration (1-2 hrs). Once the dye front (indicated by visible dye colour) reaches other end of the gel, stop electrophoresis.

4) **Gel documentation:** Gel is taken out of the tray and visualized and photographed using Gel documentation system. Molecular weight of an unknown DNA fragment is calculated by comparing relative migration of DNA fragments with that of molecular weight marker (Sambrook and Russel, 2001)

13.3 PRINCIPLE OF POLYMERASE CHAIN REACTION

Introduction

Polymerase chain reaction refers to in-vitro amplification of targeted gene. It is a rapid method for making multiple copies of specific DNA sequence. PCR essentially makes use of a short complimentary DNA oligonucleotide known as 'primer'. During *in vitro* enzymatic amplification of specified target DNA millions of copies are produced within a span of few hours resulting in amplified fragments known as 'amplicons'.

Different components required for PCR and their brief functions are as follows:

i)   **DNA:** DNA extracted from the target tissue sample containing targeted gene sequence is the most important component of PCR. Quality of DNA is crucial for getting optimum results in PCR amplification.

ii) **Primers:** They are oligonucleotides complementary to targeted gene sequence. Typical PCR will have 'forward primers' and 'reverse primers' complementary to initial and final part of targeted nucleotide sequence respectively. Primers for known gene sequence can be designed using 'Primer select' program of National Centre for Biotechnology Information database (www.ncbi.nlm.nih.gov.in).

iii) **Deoxyribo nucleotide triphosphates (dNTPs):** Four dinucleotides *viz.,* Deoxyribo Adenosine Triphosphate (dATP), Deoxyribo Cytosine Triphosphate (dCTP), Deoxyribo Guanine Triphosphate (dGTP) and Deoxy ribothymidine triphosphate (dTTP) are essential for synthesis of targeted nucleotide sequence. They are available commercially either separately or in combination.

iv) **DNA Polymerase enzyme:** Polymerase enzyme enables linking of dNTPs complementary to targeted nucleotide sequence thereby helping in synthesis and copying of targeted DNA sequence. *Taq* DNA polymerase extracted from *Thermus aquaticus* bacteria is the commonly used enzyme in PCR reactions.

v)   **Water:** Sterile de-ionized, nuclease free water free from contaminants provides perfect medium for Polymerase Chain Reaction.

vi) **10X Assay buffer:** Assay buffer creates chemical environment for action of *Taq* Polymerase hence crucial for PCR reaction. It is supplied along with *Taq* Polymerse enzyme by commercial firms. Normally the buffer contains KCl (500 mM), Tris HCl (100 mM, pH 9), $MgCl_2$ (15 mM) and gelatin (0.1% w/v).

PCR is undertaken in an equipment called 'Thermocycler'. PCR ingredients are mixed in required proportion and placed in thermocycler for reaction. Essentially PCR involves following five steps:

Step 1: Initial denaturation: PCR reaction begins with heating of reaction mix at high temperature for long duration. Normally reaction mix is heated up to 95°C for five to ten minutes to enable denaturation or uncoiling of DNA strands.

Step 2: Denaturation: This is the first step of PCR cycling in which reaction mix is heated at 95°C for short duration for unwinding of DNA double strands.

Step 3: Annealing: In this step, primers anneal to complementary strand on the uncoiled DNA strands. Normally annealing temperature varies depending on primer content and needs to be standardized for each set of primers.

Step 4: Extension: In extension step, the DNA polymerase binds to the template DNA at primer binding sites and adds new dNTPs leading to *in vitro* synthesis of DNA. Usually it is undertaken at 72°C.

PCR steps from 2 to 4 are repeated to about 30 to 45 times depending on reaction requirements.

Step 5: Final extension: Once PCR cycles are completed, final extension need to be done at 72°C for 5 to 15 minutes.

Reaction mix at the end of PCR reaction will be run on agarose gel electrophoresis for analysis of amplification results (Nagappa *et al.*, 2012).

13.4 SPECIES IDENTIFICATION OF MEAT BY FORENSICALLY IMPORTANT NUCLEOTIDE SEQUENCING

Results of forensic authentication of meat species are often subjected to legal scrutiny. Hence, techniques used for speciation must be accurate, reliable and unambiguous. Nucleotide sequencing of mitochondrial genes and alignment with the available sequences on the nucleotide database can unambiguously confirm the species origin of meat. This chapter provides Forensically Important Nucleotide Sequencing (FINS) of mitochondrial 12S rRNA gene.

Principle

Forensically Important Nucleotide Sequencing method involves extraction of DNA from meat, PCR amplification of partial fragment of mitochondrial 12S rRNA gene, sequencing of amplicon followed by alignment of sequence in Basic Local Alignment Search Tool (BLAST) of National Centre for Biotechnology Information (NCBI) database (www.ncbi.nlm.nih.gov).

Reagents and chemicals required

- **Primers:** Forward primer: 5′ CAA ACT GGG ATT AGA TAC CCA CTA 3′

  Reverse primer: 5′ GAG GGT GAC GGG CGG TGT GT 3

- Genomic DNA extracted from sample
- dNTPs
- *Taq* Polymerase
- Assay buffer
- Autoclaved nuclease free water

Procedure

- **PCR Reaction mix:** Amplification to be carried out in 0.2 ml PCR tubes containing 25 µl of PCR ready mix (2X), 1 µl (15 pmol) each of forward and reverse primers, 1 µl (50 ng) of template DNA and autoclaved Milli- Q water to make a volume up to 50 µl. PCR Ready mix containing dNTPs, *Taq* Polymerase and assay buffer in required proportion and can directly be used for PCR reaction. It will be available commercially and can be procured from any reputed firm.

- **PCR schedule:** 5 min at 94°C for initial denaturation, followed by 30 cycles of amplification (45 s at 94°C, 1 min at 60°C and 1 min at 72°C) and final extension for 10 min at 72°C.

- **Electrophoresis of PCR reaction mix:** The PCR products were analyzed by electrophoresis in 1.5% agarose gel with ethidium bromide staining. Result was recorded in gel documentation system. Size of the amplicon is 456 bp. Results of electrophoresis is given in Figure 13.1.

- **Sequencing of amplicons:** Send PCR amplicons (about 50 µl) along with primers (about 10 µl each) to commercial sequencing facility for sequencing.

- **Analysis of sequence results:** Mitochondrial 12S RNA gene sequences of different meat animal species are aligned to calculate and divergence scores using 'Megalign' program of DNA STAR software. Alignment and divergence scores are given in Table 13.1. Aligning of the nucleotide sequence of mt 12S rRNA gene of reference sample with that of sequences of database will show highest score for corresponding species hence it can clearly and unambiguously identify species of meat.

L B Bu M C P Mi

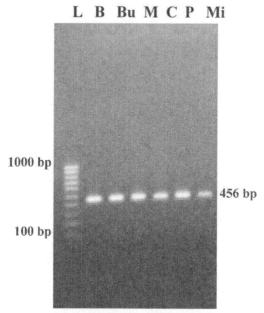

Fig. 13.1. Polymerase chain reaction amplification of mitochondrial Cytochrome B gene of meat samples run on 1.5 % agarose gel; L: 100 bp ladder; B: Beef; Bu: Buffalo; M: Mutton; C: Chevon; P: Pig; Mi: Mithun

PCR amplification, sequencing of amplicon and alignment involved in FINS can authentically detect species of meat and is applicable to varied species of meat animals. Hence, FINS can be a handy technique to food analysts for authentication of species meat (Grish *et al.*, 2004).

13.5   SPECIES AUTHENTICATION OF MEAT BY POLYMERASE CHAIN REACTION - RESTRICTION FRAGMENT LENGTH POLYMORPHISM

Introduction

Polymerase Chain Reaction – Restriction Fragment Length Polymorphism (PCR – RFLP) is a simple method for identification of species of meat and it obviates the requirement of sequencing of amplicons which is costly and time consuming. This is one of the most commonly used method for species identification of meat. This method can be used for detecting adulteration of meat and can also be used for processed meat products. In this chapter, brief description of steps involved in identification of meat using PCR RFLP of mitochondrial 12S rRNA gene is given.

Table 13.1: Nucleotide sequence similarity (upper triangle, %) and divergences (lower triangle, %) of mitochondrial cytochrome B gene of different meat animal species.

	Bos indicus	Bos grunniens	Bos frontalis	Bubalus bubalis	Bubalus depressicornis	Ovis aries	Capra hircus	Bos Sus scrofa
Bos indicus	–	94.7	96.5	90	89.5	87	89.9	82.7
Bos grunniens	3.9	–	96.5	89	89	87.7	88.9	84
Bos frontalis	3.6	2.3	–	91.5	90.7	87.5	90.2	85
Bubalus bubalis	7.5	7.7	6.9	–	98.5	88.2	89.4	82
Bubalus depressicornis	8.3	8.1	7.7	1.5	–	87.5	88.7	82.2
Ovis aries	9.5	11	8.9	8.6	9.5	–	93.5	80.5
Capra hircus	8.4	9.5	8.1	8	8.9	4.7	–	80.9
Sus scrofa	14.6	14.9	15.2	13.9	14.2	15.2	15.8	–

Principle

PCR RFLP method uses variation in nucleotide sequences in particular gene of different species. When long sequence of DNA is sliced at specific sites using REs, species-specific fingerprint is obtained. Based on unique fingerprint, animal species are identified and differentiated. PCR - RFLP involves extraction of DNA from meat using standard protocol, PCR amplification of mitochondrial Cytochrome B gene followed by restriction digestion of amplicons using Restriction Enzymes (REs). The REs are enzymes that cleave double stranded DNA at specific sites. Brief description of steps involved species identification using PCR RFLP is given below.

Chemicals and reagents required

- **Primers:** Forward primer: 5′ CAA ACT GGG ATT AGA TAC CCA CTA 3′;

  Reverse primer: 5′ GAG GGT GAC GGG CGG TGT GT 3′
- Genomic DNA extracted from sample
- dNTPs
- *Taq* Polymerase
- Assay buffer
- Autoclaved nuclease free triple distilled water
- Restriction enzymes

Procedure

- **PCR amplification:** Amplification was carried out in 0.2 ml PCR tubes containing 5 µl of 10x PCR buffer (100 mM Tris–HCl, pH 9.0, 15 mM MgCl2, 500 mM KCl and 0.1% gelatin), 1 µl of 10 mM dNTP mix, 1 µl (20 pmol) each of forward and reverse primers, 1 U of Taq DNA polymerase, 50 ng of purified DNA and autoclaved Milli-Q water to make a volume up to 50 µl.
- **PCR schedule:** 5 min at 94°C for initial denaturation, followed by 30 cycles of amplification (45 s at 94°C, 45 s at 60°C and 1 min at 72°C) and final extension for 10 min at 72°C.
- **Electrophoresis:** The PCR products were analyzed by electrophoresis in 1% agarose gel with ethidium bromide staining for confirmation of PCR.
- **Restriction digestion of PCR product:** PCR amplicons of the mitochondrial 12S rRNA gene were subjected to restriction enzyme digestion with suitable

restriction enzymes according to the suppliers instructions. Briefly, enzyme-buffer mix was prepared by mixing 2 μl of restriction enzyme with 8 μl of the respective buffer. Reaction mix was prepared by mixing 10 μl PCR product with 2 μl of enzyme buffer mix. Volume was made up to 20 μl with autoclaved MilliQ water and incubated overnight at 37°C.

- **Electrophoresis:** Digested product was visualized by electrophoresis in 2% agarose gel along with 100 bp ladder.

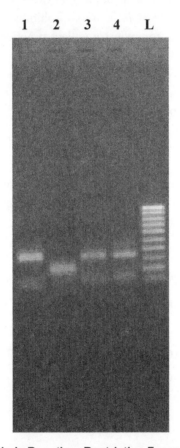

Fig. 13.2. Polymerase Chain Reaction -Restriction Fragment Length Polymorphism (PCR-RFLP) of Mitochondrial 12S rRNA Gene
Lane 1: Beef + *Alu* I **(359 + 97bp); 2: Buffalo meat +** *Hha* I **(246 + 210 bp);**
3: Mutton + *Apo* I **(339 + 117bp); 4: Chevon +** *Bsp*TI **(323 + 133 bp);**
L: 100 bp DNA ladder

Results of species identification by PCR RFLP of mitochondrial Cytochrome B gene is highly reproducible (Fig. 13.2) and can identify species even in cooked meat and could detect adulteration of beef up to 10 % in buffalo meat. It is also cheaper as compared to FINS as it does not require expensive sequencing for interpretation of results. Hence, it is an ideal method for identification of beef and detection of beef adulteration (Grish *et al.*, 2007).

13.6 AUTHENTICATION OF BUFFALO MEAT BY SPECIES SPECIFIC POLYMERASE CHAIN REACTION

Buffalo (*Bubalus bubalis*) meat is a major item of export from India but export of beef *i.e.* meat from cattle (*Bos indicus*) is prohibited. Also, misrepresentation of beef for buffalo meat is a common fraudulent practice because of prohibition on cow slaughter in most states of India. Hence, food analysts require precise identification techniques to implement such regulations. This chapter provides description of basic steps involved in identification of buffalo meat by species specific PCR targeting mitochondrial D loop region.

Principle

Species specific PCR involves amplification of specific region of the gene by utilizing set of specific primers targeting a species. It takes lesser assay time as it does not involve any post reaction processing except electrophoresis. Primers yield amplicon only in the species for which they have been designed and no amplification is expected in other species.

Chemicals and Reagents

- **Primers:** Forward- 5' - TGA CCC TAC TAC TCC GAATGG -3'
  Reverse-5' - GCT GAG TCC AAG CAT CCC -3'
- Genomic DNA extracted from sample
- dNTPs
- *Taq* Polymerase
- Assay buffer
- Autoclaved nuclease free triple distilled water

Procedure

- **PCR reaction mix:** PCR amplification to be carried out in 0.2 ml PCR tubes containing 5 µl o f 10X PCR buffer (100 mM Tris – HCl, pH 9.0, 15 mM MgCl2, 500 mM KCl and 0.1% gelatin), 1 µl of 10 mM dNTP mix, 1 µl (15 pmol) each of forward and reverse primers, 1 U of Taq DNA polymerase, 50 ng of purified DNA and autoclaved Mili- Q water to make a volume up to 50 µl.

- **PCR schedule:** 5 min at 94°C for initial denaturation, followed by 30 cycles of amplification (45 s at 94°C, 45 s at 60°C and 1 min at 72°C) and final extension for 10 min at 72°C.

- **Electrophoresis:** The PCR products were analyzed by electrophoresis in 2% agarose gel with ethidium bromide staining. Result is depicted in Fig. 13.3.

Results and interpretation

Amplicon of size 482 bp can be found in buffalo whereas no amplification can be found in other related species *viz.*, cattle, sheep and goat as shown in figure.

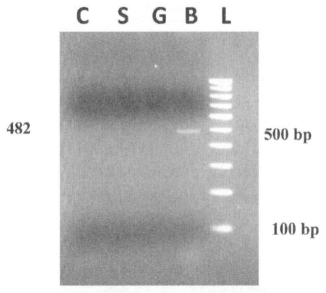

Fig. 13.3. Buffalo specific PCR targeting mitochondrial D loop gene
C: Cattle; S: Sheep; G: Goat; B: Buffalo; L: 100 bp DNA Ladder

Species specific PCR is a single step method which obviates the requirement of sequencing as in FINS and restriction enzyme digestion of amplicons as in RFLP. Hence, it is a quick and economical method for identification of buffalo meat which will be handy to experts involved in authentication of export consignments (Grish *et al.*, 2013).

13.7 AUTHENTICATION OF SHEEP MEAT (MUTTON) BY SPECIES SPECIFIC POLYMERASE CHAIN REACTION

Misrepresentation of mutton for chevon is a common fraudulent practice for economical reasons in different states of India. To control any misrepresentations affecting consumer sentiments authentic identification techniques are required. This chapter provides brief description of steps involved in authentication of mutton using species specific polymerase chain reaction targeting mitochondrial D loop region.

Principle

Species specific PCR involves amplification of specific region of the gene by utilizing set of specific primers targeting a species. It takes lesser assay time as it does not involve any post reaction processing except electrophoresis. Primers yield amplicon only in the species for which they have been designed and no amplification is expected in other species. Methodology involves extraction of DNA from meat samples, PCR amplification using set of novel primers designed targeting mitochondrial D loop gene followed by agarose gel electrophoresis.

Chemicals and reagents

- Primers: Forward-5′ -AAA AGC ACA ACC ATC CAC CC-3′

  Reverse-5′ -GTC AAG CAG TTC AAT GCC CTA T-3′
- Genomic DNA extracted from sample
- dNTPs
- *Taq* Polymerase
- Assay buffer
- Autoclaved nuclease free triple distilled water

Procedure

- **PCR Reaction mix:** PCR amplification to be carried out in 0.2 ml PCR tubes containing 5 μl o f 10X PCR buffer (100 mM Tris – HCl, pH 9.0, 15 mM MgCl2, 500 mM KCl and 0.1% gelatin), 1 μl of 10 mM dNTP mix, 1 μl (15 pmol) each of forward and reverse primers, 1 U of Taq DNA polymerase, 50 ng of purified DNA and autoclaved Mili- Q water to make a volume up to 50 μl.

- **PCR Reaction schedule:** 5 min at 94°C for initial denaturation, followed by 30 cycles of amplification (1 min at 94°C, 1 min at 56°C and 1 min at 72°C) and final extension for 5 min at 72°C.

- **Electrophoresis:** The PCR products can be analyzed by electrophoresis in 2% agarose gel with ethidium bromide staining.

Results and interpretation

PCR yields an amplicon of 425 bp. However non specific amplicon of about 350 bp can be found which will be repetitive. With sheep DNA designed primers will yield amplicons of size 425 bp and 350 bp consistently (Figure 13.4). Such a phenomenon is not uncommon in sheep owing to 'heteroplasmy' *i.e.* occurrence of more than two types (copy number) of mitochondria in the cells. No cross amplification can be found in cattle, goat and buffalo DNA.

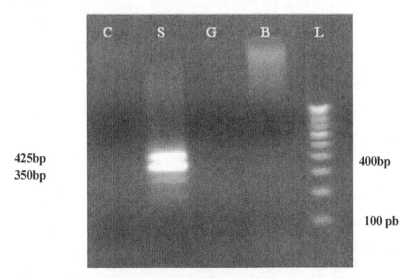

Fig. 13.4: Sheep specific PCR amplification run on 2 % agarose gel.
C: Cattle;S: Sheep; G: Goat; B: Buffalo

Species specific PCR does not require any post PCR manipulations like restriction enzyme digestion, sequencing of amplicons etc. Hence, species specific PCR targeting mitochondrial D loop gene is an ideal method for authentication mutton and can be used by food analysts for forensic analysis when the terms of reference is limited. Positive amplification indicates sample as mutton and negative amplification indicates sample is not mutton. However, method will not be able identify which species it belongs to if referred meat sample is not mutton (Grish *et al.*, 2016).

13.8 AUTHENTICATION OF GOAT MEAT (CHEVON) BY SPECIES SPECIFIC POLYMERASE CHAIN REACTION

Due to its leanness, chevon is a preferred meat by many consumers. Some meat vendors indulge in adulteration of chevon with cheaper meats for financial gains. To curb such malpractices authentic detection tools are required. This chapter gives brief description on steps involved in authentication of chevon.

Principle

Species specific PCR involves amplification of specific region of the gene by utilizing set of specific primers targeting a species. It takes lesser assay time as it does not involve any post reaction processing except electrophoresis. Primers yield amplicon only in the species for which they have been designed and no amplification is expected in other species. Methodology involves extraction of DNA from meat samples, PCR amplification using set of novel primers designed targeting mitochondrial D loop gene followed by agarose gel electrophoresis.

Reagents and chemicals

- **Primers:** Forward-5′ -GGA AGA AGC CAT AGC CTC AC-3′
  Reverse-5′ -AGT TGG GTT AGG ATT GGG AT-3′
- Genomic DNA extracted from sample
- dNTPs
- *Taq* Polymerase
- Assay buffer
- Autoclaved nuclease free triple distilled water

Procedure

- **PCR reaction mix:** PCR amplification to be carried out in 0.2 ml PCR tubes containing 5 µl o f 10X PCR buffer (100 mM Tris – HCl, pH 9.0, 15 mM MgCl2, 500 mM KCl and 0.1% gelatin), 1 µl of 10 mM dNTP mix, 1 µl (15 pmol) each of forward and reverse primers, 1 U of Taq DNA polymerase, 50 ng of purified DNA and autoclaved Mili- Q water to make a volume up to 50 µl.
- **PCR reaction schedule:** 5 min at 94°C for initial denaturation, followed by 30 cycles of amplification (1min at 94°C, 1 min at 56°C and 1 min at 72°C) and final extension for 5 min at 72°C.

- **Electrophoresis:** The PCR products were analyzed by electrophoresis in 2% agarose gel with ethidium bromide staining.

Results and interpretation

PCR will yield an amplicon of size 229 bp in goat DNA. Result is depicted in Figure 13.5. Species specific PCR is a simple one step method for authentication of chevon. No amplification can be seen in other meat animal species.

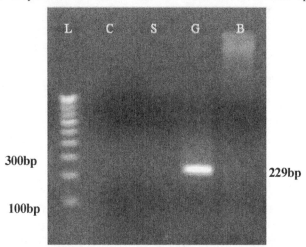

Fig. 13.5: Goat specific polymerase chain reaction targeting mitochondrial D loop region
L: 100 bp DNA Ladder; C: Cattle; S: Sheep; G: Goat; B: Buffalo

13.9 MEAT SPECIES IDENTIFICATION BY RANDOMLY AMPLIFIED POLYMORPHIC DNA - POLYMERASE CHAIN REACTION

Randomly Amplified Polymorphic DNA Polymerase Chain Reaction (RAPD - PCR) involves PCR amplification using randomly designed primers without targeting any specific gene. As it does not require prior knowledge of gene sequences assay is simple to design. Single step of PCR using random primers can identify different meat animal species. However, RAPD PCR has draw backs of non reproducibility and hence it is not a commonly used method for species identification of meat.

Principle

In RAPD, several non-specific primers are used to amplify many amplicons simultaneously and resultant banding pattern (obtained upon gel electrophoresis) is interpreted to identify animal species. Such band patterns vary between species due to nucleotide sequence variation. A reproducible band pattern observed after gel electrophoresis is referred as 'RAPD fingerprint'. Target DNA sequences yield reproducible fingerprints and help in animal species identification and such sequences are known as 'RAPD markers'. Animal species can be identified and discriminated using such unique species-specific fingerprints generated by the RAPD-PCR discrimination

Reagents and chemicals

- Primers: 5' GTC GGC GCA G 3'
- Genomic DNA extracted from sample
- dNTPs
- *Taq* Polymerase
- Assay buffer
- Autoclaved nuclease free triple distilled water

Procedure

- **PCR reaction mix:** Nucleotide sequence of the primer used for RAPD PCR was. Reaction was performed in 0.2 ml capacity thin walled PCR tube by adding 5 µl of 10X Taq DNA polymerase buffer (100 mM Tris HCl, pH 9.0, 500 mM KCl, 15 mM $MgCl_2$, 0.1%W/V gelatin), 0.25 mM of dNTP mixture (dATP, dCTP, dGTP and dTTP), 20 pmol of primer, 50 ng of template DNA and 0.3 µl (1.0 units) of Taq DNA polymerase. The volume was made up to 50 µl with DNase free ultrapure distilled water. The PCR tube containing the reaction mixture was flash-spun on a micro centrifuge to get the reactants at bottom.

- **PCR reaction schedule:** Initial denaturation at 94°C for 5 min followed by 45 cycles of 1 min denaturation at 94°C, 45 sec annealing at 36°C and 1 min elongation at 72°C. It was followed by final extension at 72°C for 5 min. After the reaction, PCR products were held at 4°C.

- **Electrophoresis:** PCR products were analyzed by electrophoresing on 2.5 % agarose gel. Molecular weights of different amplicons in the RAPD profiles were calculated using Alphaimager^R software.

Results and interpretation

- Result of PCR amplification is depicted in Figure 13.6. Amplicon size found in different species are given in Table 13.2. As given in table different species gave amplicons of different sizes. Hence, method can differentiate species in single PCR process.

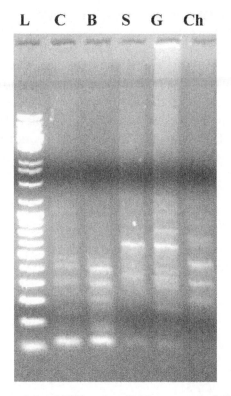

**Fig. 13.6. RAPD fingerprints of DNA extracted from meat of different species run on
2.5 % agarose gel electrophoresis**
L: DNA Ladder; C: Cattle; B: Buffalo; S: Sheep; G: Goat; Ch: Chicken

Table 13.2: Size of RAPD PCR amplicons in different meat animals

Sl No.	Species	Amplicon size (bp)
1	Cattle (*Bos indicus*)	148, 234, 363, 452
2	Buffalo (*Bubalus bubalis*)	148, 273, 345, 430
3	Sheep (*Ovis aries*)	148, 645, 389
4	Goat (*Capra hircus*)	145, 613, 777,375
5	Chicken (*Gallus gallus*)	263, 351, 467, 689

Advantage of RAPD is that single PCR process can identify different meat animal species. However, RAPD is affected by quality of DNA and reaction conditions. Sheared DNA from cooked meat samples will not give adequate result in RAPD. Repeatability of results is found to be lower as compared to other methods. Hence, it is not a commonly used method for speciation in recent years (Girish *et al.*, 2011).

13.10 IDENTIFICATION OF SEX OF MEAT BY PCR AMPLIFICATION OF AMELOGENIN XY GENE

Slaughter of female cattle (cow) is prohibited in most states of India. However, slaughter of male cattle (bullocks/bulls) is permitted with restrictions. Some states have prohibition on slaughter of cow and its progeny *i.e.* slaughter of both male and female cattle are prohibited. In pigs and sheep, meat from male animal is preferred over that of female animal. This necessitates differentiation of sex of meat. This chapter provides brief description of steps involved in identification of sex of meat.

Principle

PCR amplification of Amelogenin XY gene yields two products of size 280 and 217 bp in males whereas in females single PCR product of size 217 bp is amplified in females of meat animals this is due to deletion region in Amelogenin Y gene which is found only in male. Single amplicon indicates that the meat sample is from female while double amplicon indicate that the sample is of female.

Reagents and chemicals

- **Primers:** Forward primer 5′ - CAG CCA AAC CTC CCTCTG C - 3′
  Reverse primer 5′ - CCC GCT TGG TCT TGT CTG TTG C – 3′
- Genomic DNA extracted from sample
- dNTPs
- *Taq* Polymerase
- Assay buffer
- Autoclaved nuclease free triple distilled water

Procedure

- **PCR Reaction mix:** PCR amplification is carried out in 0.2 ml PCR tubes containing 2.5 μL of 10x PCR buffer (100 mM Tris–HCl, pH 9.0, 15 mM MgCl2, 500 mM KCl and 0.1% gelatin), 1 μL of 10 mMdNTP mix, 1 μl (20 pmol) each of forward and reverse primers, 1 U (IU?) of Taq DNA polymerase, 50 ng of purified DNA and autoclaved Milli- Q water to make volume up to 25μl.

- **PCR schedule:** 5 min at 94°C for initial denaturation, followed by 34 cycles of amplification (45 s at94°C, 45 s at 60°C and 1 min at 72°C) and final extension for 10 min at 72°C.

- **Electrophoresis:** The PCR products were analyzed by electrophoresis in 2% agarose gel with ethidium bromide staining.

Results and interpretation

PCR amplification yields single amplicon of 217 bp in female and double amplicons of size 217 and 280 bp in female which makes it easy to differentiate sex of meat (Fig. 13.7).

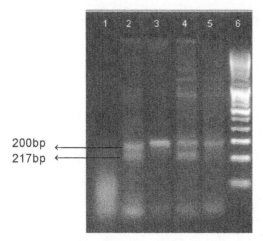

Fig. 13.7. Identification of Sex of Meat by Polymerase Chain Reaction Amplification of Amelogenin XY Gene

Line 1: Negative control; Line 2: Cattle (♂); Line 3: Cattle (♀); Line 4: Buffalo (♂);

Line 5: Buffalo (♀); Line 6: DNA Ladder

13.11 SPECIES IDENTIFICATION OF PORK BY LOOP MEDIATED ISOTHERMAL AMPLIFICATION

Adulteration of meat with pig meat is considered as an offence owing to the religious and health concerns. Pig meat and products being readily available and cheap sources, are substituted into other species meats. Scandals associated with incorporation of pig meat into other species meats have drawn considerable attention in the past. In order to comply with *halal* authentication, protect consumers' sentiments, promote fair-trade, implement prompt labeling norms, avoid allergies and prevent disease transmission from pig or pork, there is a need to correctly identify pig derived tissues.

Principle

Loop-mediated Isothermal Amplification (LAMP) is a nucleotide amplification technique that amplifies target DNA at isothermal temperatures and obviates need for thermal cyclers and post-amplification procedures of signal detection. As in PCR, LAMP assay could amplify target DNA several folds (10^9 copies in a span of an hour) under isothermal conditions without compromising specificity and having capability of the visual detection of amplified targets using specific dyes. Consequently, LAMP assay has emerged as an alternative tool to PCR based techniques for the purpose of testing food safety hazards including detection of meat adulteration.

Reagents and chemicals

- Two sets of primers (F3/B3, FIP/BIP)

  F3 – AGG CCC TAA CAC AGT CAA

  B3 – GTT ATA GGG TGT GTA GAG CAT A

  FIP - ACT GAA TAG CAC CTT GTT TGG ATT TGT AGC TGG ACT TCA TGG

  BIP - CGG GAC ATA ACG TGC GTA CAA GTT TAA TGG GGG GTA AGG.
- Extracted tissue DNA
- 10X Thermopol buffer
- Betaine
- 50 mM $MgSo_4$
- 10 mM dNTP mix
- *Bst*I Enzyme
- SYBR green I dye

Procedure

The LAMP reaction mix is prepared by adding 10X Thermopol buffer (3 μL), 5 M Betaine (3 μL), 50 mM $MgSo_4$ (3 μL), 10 mMdNTP mix (4 μL), 10 pmol F3 primer (1 μL),10 pmol B3 primer (1 μL), 40 pmol FIP primer (4 μL), 40 pmol BIP primer (4 μL) and template DNA (5 μL). The reaction mix was heated at 95 °C for 5 min for denaturation; thereafter, 8 units (1 μL) of *Bst*I enzyme was added and incubated at 65 °C for 1 h followed by heating at 80 °C for 2 min for the inactivation of *Bst*I enzyme. After the LAMP reaction, 1 μL of SYBR green I dye (1:10, 10,000X) was added for the visualization of amplification.

Results and interpretation

Positive reaction is indicated by the green colour; whereas, orange color is indicative of the negative reaction. Results can be interpreted by colour change only.

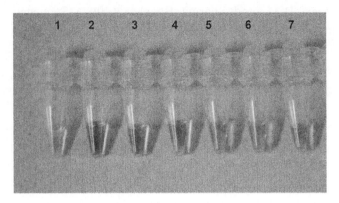

**Fig. 13.8. Visualization of LAMP amplification after addition of SYBR Green I dye
1) Pig; 2) Cattle; 3) Buffalo; 4) Sheep; 5) Goat; 6) Chicken; 7) Negative control
(Girish et al., 2020)**

References

Ennis S, Gallagher TF (1994) A PCR- based sex-determination assay in cattle based on thebovine amelogenin locus. Animal Genetics 25:425-427.

Grish, P.S. Vaithiyanathan, S., Nagappa Karabasanavar and Shailish Bagoli (2016). Authentication sheep (ovis aries) and goat (capra hircus) meat using species specific polymerase chain reaction. Indian Journal of Animal Science, 86(10), 1172-1175.

Girish, P. S., Haunshi, S., Vaithiyanathan, S., Rajitha, R. & Ramakrishna, C. (2013) A rapid method for authentication of Buffalo (*Bubalus bubalis*) meat by alkaline lysis method of DNA extraction and species specific polymerase chain reaction. Journal of Food Science & Technology, 50 (1):141–146

Girish, P. S., Haunshi, S., Vaithiyanathan, S., Rajitha, R. & Ramakrishna, C. (2013) A rapid method for authentication of Buffalo (*Bubalus bubalis*) meat by alkaline lysis method of DNA extraction and species specific polymerase chain reaction. Journal of Food Science & Technology, 50 (1):141–146

Girish, P.S., A.R. Sen S. Vaithiyanathan, Y. Babji, R. Rajitha and M. Muthukumar (2011) Meat species identification by randomly amplified polymosphic DNA - Polymerase Chain Reaction. Journal of Meat Science, 7(01), 28-31.

Girish, P.S., Anjaneyulu, A.S.R., Viswas, K.N., Shivakumar, B.M., Anand, M., Patel, M. and Sharma, B. (2005) Meat species identification by Polymerase chain reaction – restriction fragment length polymorphism (PCR-RFLP) of mitochondrial 12S rRNA gene. Meat Science, 70:107-112

Girish, P.S., Anjaneyulu, A.S.R., Viswas, K.N., Anand, M., Rajkumar, N., Shivakumar, B.M. and Sharma Bhaskar (2004) Sequence analysis of mitochondrial 12S rRNA gene can identify meat species. Meat Science, 66: 551-556.

Girish, P.S., Barbuddhe, S. B., Aparana Kumari, Deepak B. Rawool, Nagappa S. Karabasanavar, M. Muthukumar and S. Vaithiyanathan (2020) Rapid detection of pork using alkaline lysis-Loop Mediated Isothermal Amplification (AL-LAMP) te chnique. Food Control, 110, 107015.

Nagappa Shivaning Karabasanavar, Girish Patil, S., Santosh Haunshi & Viswas, K. N. (2012) DNA based animal species identification. Hind Publisher, Hyderabad.

Nagappa Shivaning Karabasanavar, Girish Patil, S., Santosh Haunshi & Viswas, K. N. (2012) DNA based animal species identification. Hind Publisher, Hyderabad, 165 pages.

Nagappa Shivaning Karabasanavar, Girish Patil, S., Santosh Haunshi & Viswas, K. N. (2012) DNA based animal species identification. Hind Publisher, Hyderabad.

Sambrook, J., & Russell, D. W. (2001). Molecular cloning—a laboratory manual. Cold Spring Harbour Laboratory Press, third edition: New York.

PROTEOMIC TOOLS FOR MEAT QUALITY EVALUATION

14.1 EXTRACTION OF TOTAL, MYOFIBRILLAR AND SARCOPLASMIC PROTEINS FROM MEAT AND MEAT PRODUCTS

14.1.1 Total protein extraction

The 100 mg of meat sample (raw/cooked) was mixed with 1mL of extraction buffer (7 M Urea, 2 M Thiourea, 4% w/v CHAPS, 2% Carrier ampholyte pH 4–7, 40 mM DTT and Protease inhibitor) using a BeadBug™ microtube homogenizer (Benchmarck scientific). Homogenated for 2 x 30 sec of 4000 cycles per minute. Hold the homogenized samples at room temperature for 1 hour with alternative vortexing/ shaking. After this samples were centrifuged using refrigerated ultracentrifuge (Eppendorf centrifuge 5430R, Germany) at 10000 rcf for 1hour at 10°C. The supernatant was collected (avoid fat layer) and stored at -80°C till further analysis (Montowska and Pospiech, 2012). Different extraction steps were presented below in Figure 14.1.

Extraction buffer: 7 M Urea, 2 M Thiourea, 4% w/v CHAPS, 2% Carrier ampholyte pH 4–7, 40 mM DTT and Protease inhibitor

0.1 g meat + 1mL extraction buffer
↓   Homogenization (bead homogenizer)
Alternate vortex/shaking for 1hr at room temperature
↓
Centrifuge at 10 000 rcf for 1 h at 10°C
↓
Collect supernatant, avoid fat layer and store at "80°C

Fig. 14.1. Extraction of total meat proteins

14.1.2 MYOFIBRILLAR AND SARCOPLASMIC PROTEIN EXTRACTION

One gram of raw/cooked meat sample was mixed with 10mL of buffer A (50mM Tris-HcL, pH 8.0) and homogenized using tissue homogenizer. After brief vortex for 2 minutes, centrifuged at 10000 rcf for 20 minutes at 4°C. Supernatant was collected and designated as sarcoplasmic protein extract. Remaining sediment was mixed with 10mL of buffer B (50mM Tris-HcL, pH 8.0, containing 6M Urea, 1M Thiourea), vortexed for 10 minutes to ensure proper mixing followed by centrifugation at 10000 rcf for 10 minutes at 4°C. Supernatant was collected and filtered through Whatman filter paper No. 1, and designated as myofibrillar protein extract. Extracted protein samples were stored at -80°C, till further analysis (Sentandreu *et al.*, 2010). Different protein extraction steps are indicated in Figure 14.2.

Buffer A: 50 mM Tris-Hcl buffer, pH 8.0

Buffer B: 50 mM Tris-Hcl buffer, pH 8.0 containing 6 M Urea and 1 M Thiourea

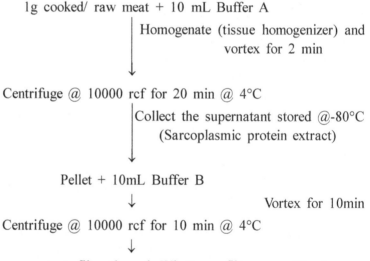

1g cooked/ raw meat + 10 mL Buffer A

Homogenate (tissue homogenizer) and vortex for 2 min

Centrifuge @ 10000 rcf for 20 min @ 4°C

Collect the supernatant stored @-80°C (Sarcoplasmic protein extract)

Pellet + 10mL Buffer B

Vortex for 10min

Centrifuge @ 10000 rcf for 10 min @ 4°C

Collect the supernatant, filter through Whattman filter paper No. 1 (Myofibrillar protein extract) stored @-80°C

Fig. 14.2. Extraction of sarcoplasmic and myofibrillar meat proteins

14.2 PURIFICATION AND QUANTIFICATION OF MEAT PROTEIN EXTRACT

14.2.1 TCA/Acetone precipitation method

Equal volumes of extracted protein sample and 50% TCA containing 20 mM DTT were mixed and incubated on ice with intermittent vortexing for 1hour. The mixture was centrifuged at 16000 rcf for 10minutes and the supernatant was decanted. The 500 µl of ice cold 100% acetone containing 20 mM DTT was added and mixed with pellet by vortexing and inversion for 15-30 minutes in ice box. Centrifuge the mixture at 16000 rcf for 10 minutes and supernatant was decanted and the acetone wash was repeated twice. Acetone was discarded and the protein pellet was dried for 5 min. The protein pellet was dissolved in Rehydration buffer (7M Urea, 2M Thiourea and 2%w/v CHAPS) and stored at -80°C (Chen *et al.*, 2005). Different protein purification steps were presented below in Figure 14.3.

Equal volume of 50% TCA + 20 mM DTT with protein extract
↓
Keep on ice for 1 hour
↓
Centrifuge at 16,000 rcf for 10 min @ 4°C
↓
Remove supernatant to obtain protein pellet
↓
Wash with 500 µl acetone (acetone + 20 mM DTT)
↓
Centrifuge for 10 min at 16000 rcf @ 4°C
↓
Discard supernatant and dry to get pure protein pellet
Dissolve in rehydration buffer

Fig. 14.3. Flow chart showing protein purification using TCA: acetone precipitation

14.2.2 Purification using 2-DE clean-up kit

Commercially available 2-D clean-up kit (GE Healthcare) was tested for its efficacy to precipitate the impurities in protein extract. The following manufacturer's guidelines were used to ensure purification of buffalo and sheep meat protein extracts.

Procedure

● Add 300 µl precipitant to 1-100 µl sample (containing 1-100 µg protein) and vortex. Incubate on ice for 15 min.

● Add 300 µl co-precipitant and mix.

● Centrifuge at 12000 rcf.

- Remove supernatant, without disturbing the pellet layer and add 40 μl of co-precipitant and keep on ice for 5 min.
- Centrifuge at 12000 rcf
- Add 25 μl of distilled water and disperse pellet by vortexing.
- Add 1 mL chilled, wash buffer and 5 μl wash additive. Incubate for 30 min. Vortex for 20-30 sec for every 10 min.
- Centrifuge at 12000 rcf.
- Remove and discard the supernatant followed by drying of pellet.

14.3 PROTEIN QUANTIFICATION

14.3.1 Protein estimation by standard BSA – BCA method

1. 50 uL of (1:1000) diluted protein is added with 1 mL of working reagent containing BCA solution and $CuSo_4$ in the ratio of 1:50 i.e., 1mL BCA solution with 20uL $CuSo_4$,

2. Samples are incubated at 37°C for 30 min in water bath and readings are taken at 562 nm with the help of UV- VISIBLE spectrophotometer.

3. Standard BSA curve is obtained from appropriate dilution from 2mg/mL stock BSA solution that ranges from 5ug/ml to 600ug/ml.

Principle: The Pierce BCA Protein Assay is a detergent-compatible formulation based on bicinchoninic acid (BCA) for the colorimetric detection and quantization of total protein. This method combines the well-known reduction of Cu^{+2} to Cu^{+1} by protein in an alkaline medium (the biuret reaction) with the highly sensitive and selective colorimetric detection of the cuprous cation (Cu^{+1}) using a unique reagent containing bicinchoninic acid. The purple-colored reaction product of this assay is formed by the chelation of two molecules of BCA with one cuprous ion. This water-soluble complex exhibits a strong absorbance at 562nm that is linear with increasing protein concentrations over a broad working range of 20μg/ml to 2,000μg/ml. The BCA method is not a true end-point method – the final colour continues to develop but, following incubation, the rate of colour development is slowed sufficiently to allow large numbers of samples to be done in a single run. The macromolecular structure of protein, the number of peptide bonds and the presence of four amino acids (cysteine, cystine, tryptophan and tyrosine) are reported to be responsible for color formation with BCA.

1. Protein (peptide bonds) + Cu^{+2} tetradentate-Cu^{+1} complex

2. Cu^{+1} + 2 Bicinchoninic Acid (BCA) BCA-Cu^{+1} complex (purple coloured, read at A562)

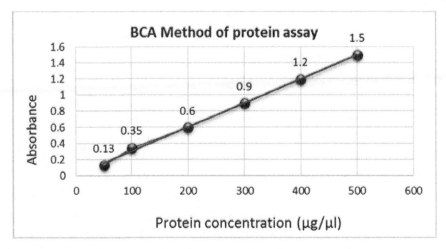

Fig. 14.4. Standard curve for protein quantification by BCA method

14.3.2 Protein estimation by 2-D Quant kit

Table 14.1. Prepare a standard curve according to the fallowing

Standard solution	1	2	3	4	5	6
2 mg/ml BSA	0 µl	5 µl	10 µl	15 µl	20 µl	25 µl
Protein quantity	0 µg	10 µg	20 µg	30 µg	40 µg	50 µg

1. Set up six tubes and add standard solution according to Table: Tube 1 is the assay blank, which contains no protein.

2. Prepare tubes containing 1–50 µl of the sample to be assayed. Duplicates are recommended. The useful range of the assay is 0.5–50 µg and it is also recommended that more than one sample volume or dilution be assayed for each sample to ensure that the assay falls within this range.

3. Add 500 µl precipitant to each tube (including the standard curve tubes). Vortex briefly and incubate the tubes 2–3 min at room temperature.

4. Add 500µl co-precipitant to each tube and mix briefly by vortexing or inversion.

5. Centrifuge the tubes at a minimum of 10 000 × g for 5 min and this sediment the protein.

6. Remove the tubes from the centrifuge as soon as centrifugation is complete. A small pellet should be visible. Decant the supernatants. Proceed rapidly to the next step to avoid resuspension or dispersion of the pellets.

7. Carefully reposition the tubes in the microcentrifuge as before, with the cap-hinge and pellet facing outward. Centrifuge the tubes again to bring any

remaining liquid to the bottom of the tube. A brief pulse is sufficient. Use a micropipette to remove the remaining supernatant. There should be no visible liquid remaining in the tubes.

8. Add 100 µl of copper solution and 400 µl of distilled or de-ionized water to each tube and Vortex briefly to dissolve precipitated protein.

9. Add 1 ml of working color reagent to each tube (See "Preliminary preparations" for preparing the working color reagent). Ensure instantaneous mixing by introducing the reagent as rapidly as possible. Mix by inversion and Incubate at room temperature for 15–20 min.

10. Read the absorbance of each sample and standard at 480 nm using water as the reference. The absorbance should be read within 40 min of the addition of working color reagent (step 9). Note: Unlike most protein assays, the absorbance of the assay solution decreases with increasing protein concentration. Do not subtract the blank reading from the sample reading or use the assay blank as the reference.

11. Generate a standard curve by plotting the absorbance of the standards against the quantity of protein. Use this standard curve to determine the protein concentration of the samples

12. For unknown protein sample, take 5ul of each protein extract in duplicates fallowing the same principle as above.

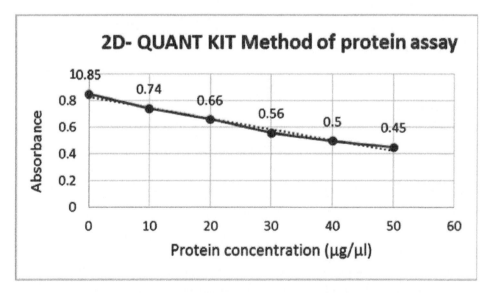

Fig. 14.5. Standard curve for protein quantification by 2D quant kit

14.4 SODIUM DODECYL SULPHATE–POLYACRYLAMIDE GEL ELECTROPHORESIS (SDS-PAGE)

Sample buffer solution (5X conc.): 0.5 M Tris-HCl pH 6.8, 50% v/v glycerol, 10% w/v SDS, 0.2 M DTT and 0.05% bromophenol blue.

Sodium dodecyl sulphate -polyacrylamide gel electrophoresis (SDS-PAGE) was carried out using the method of Laemli (1970) with midi-electrophoresis apparatus (GE Healthcare, Uppsala, Sweeden, Model: SE-600 Ruby). Composition of stacking and resolving gel given in Table 1. About 50 µl of total protein sample (containing 50 µg of protein) mixed 5X sample buffer heated to 95°C for 5 minutes was used for loading the 12% gel. Electrophoresis was performed at a constant voltage mode of 100 V/slab at 60 mA for 7-8 h or until the sample reached the lower end of the gel. Lastly the gel is *stained* (0.2% coommossic blue, 40% methanol, and 10 % Acetic acid) and distained (40% methanol, 10 % Acetic acid) for visualization of protein and gel photograph is taken and stored.

Table 14.2. Composition of different gel components for SDS-PAGE

Ingredient	Stacking gel (mL)	Resolving gel (mL)
Dist. Water	1.4	3.4
Tris-buffer (pH)	0.625 (6.8)	2.5 (8.8)
30% Acrylamide	0.335	4
10%SDS	25 µl	100 µl
10% APS	25 µl	100 µl
TEMED	3 µl	9 µl

14.5 ISOELECTRIC FOCUSSING BY USING AGILENT 3100 OFFGEL FRACTIONATOR

Assembling OFFGEL Fractionator

1. Place the Tray in the orientation shown with fixed electrode side on the left and handle the tray on right and remove protective backing of the IPG strip,

2. Place the IPG strip in the tray with gel side up with handle on the right and fixed electrode slots on the left,

3. Pull the strip as far as left as possible until the strip touches the left edge of the tray (low pH side (anode) is marked with + must be left side while fixed electrode will be attached),

4. Place left and right side of the frame and press them until it snaps in,

Procedure for OFFGEL Fractionation of proteins

1. Pipette 40uL of IPG strip rehydration solution in to each well of the 24cm IPG strips see that it not to touch the gel and gently tap the tray into the desk to ensure that solution reaches the gel,

2. Then place one of the two of each electrode pad wet with IPG rehydration solution on either side of the tray

3. Wait 15 min to allow IPG gel to swell,

4. After swelling, load 150ul of prepared (protein or peptide) OFFGEl sample (OFFGEL stock solution along with the sample) into each well,

5. Then, place the cover seal over the frame and press down gently on each well to proper fit, care is taken to not to move the main frame in the tray,

6. Re-apply 10ul dH_2O onto each electrode pads at each of the gel ends, care is taken to not to move the pads,

7. Place the tray on the instrument platform,

8. Finally pipette cover fluid (mineral oil) onto the gel strip ends in several steps,

9. For 24 well frames, pipette 200uL cover fluid onto the anode end (fixed electrode) of the IPG strip, pipette 400uL cover fluid at the cathode side (movable electrode), after 1min, reapply an additional 200uL cover fluid to both ends of the IPG strips, then after 3 min, add additional 200uL cover fluid at the anode end (fixed electrode) of the IPG strip.

NOTE:

1. Cover fluid should not exceed higher than ½ the height of the tray grooves.

2. Change the electrode pads for each 24hours of the operation.

3. Fix gently and carefully onto tray platform and finally run the protocol.

Table 14.3. OFFGEL stock solution (1.25X) composition and preparation and protein
OFFGEL STOCK SOLUTION (1.25X)

Tube	Amount	Comment
Thiourea	Entire tube	Rinse with dH_2O if needed
DTT	Entire tube	Rinse with dH_2O if needed
Glycerol solution	6mL	
OFFGEL Buffer	600uL	Choose the appropriate ampholytes depending on the IPG strip used- P^H 3-10 or 4-7

Table 14.4. Protein IPG Strip Rehydration solution

Subject	24cm strip/ 24 well frame	12cm strip or 12 well frame
Protein OFFGEL Stock solution (1.25x)	0.96 mL	0.56 mL
dH$_2$0	0.24 mL	0.14mL
Total solution	1.2 mL	0.7 mL

Table 14.5. OFFGEL Protein Sample preparation

Subject	24cm strip/ 24 well frame	12cm strip or 12 well frame
Protein OFFGEL Stock solution (1.25x)	2.88mL	1.44 mL
Protein sample	0.72 mL	0.36mL
Total solution	1.2 mL	0.7 mL

NOTE: the final salt concentration in the diluted sample should not exceed 10 mM (Naveena *et al.*, 2017)

14.6 TWO-DIMENSIONAL GEL ELECTROPHORESIS (2-DE)

14.6.1 1st Dimension- Isoelectric focusing (IEF)

IPG strips: 13cm Immobilin^{TM} Drystrip, 3-10 pH (GE Healthcare, Bio-Sciences).

Rehydration buffer: 7M Urea, 2M Thiourea, 2% CHAPS, 50mM DTT and 0.5-2% ampholytes

Sample preparation: The TCA/Acetone precipitated protein samples in Rehydration buffer (without DTT and Ampholytes) is quantified using 2D Quant kit.

Procedure: Approximately 450-550 μg of TCA/Acetone precipitated protein sample were mixed with 250 μl rehydration buffer for 13cm IPG strip (Immobilin^{TM} Drystrip, GE Health care, Uppsala, Sweden) with 50mM DTT and 2% of 3-10 pH Ampholytes. Rehydration buffer along with sample was smeared on Ettan IPG strip holder (GE Healthcare, Bio-Sciences) and the IPG strip gel side facing to the buffer was placed over the sample for passive rehydration (typically 12-16 hrs.) at room temperature. Later, IPG strips were focused using an Ettan IPGphor 3 System (GE Healthcare, Bio-Sciences) with a ceramic manifold as per the protocol mentioned in Table 14.6.

Table 14.6. IEF time-current combination for Ettan IPGphor 3 System

Step No	Current type	Voltage (V)	Final Voltage (Vhrs)
1	Step	300	300
2	Step	500	500
3	Gradient	1000	800
4	Gradient	4000	11300
5	Gradient	8000	14100
6	Step	8000	7400
7	Hold. Step	500	20:00(HH:MM)

14.6.2 2nd Dimension SDS-PAGE

Equilibration buffer I: 6M Urea, 2% SDS, 50mM Tris Hcl (pH 8.8), 30% glycerol, 0.0025% Bromophenol blue and 1% w/v DTT.

Equilibration buffer II: 6M Urea, 2% SDS, 50mM Tris Hcl (pH 8.8), 30% glycerol, 0.0025% Bromophenol blue and 2.5% w/v Iodoacetamide (IAA). The focused IPG strips were equilibrated using equilibration buffer I and equilibration buffer II for 15 minutes each. After equilibration, proteins were separated in the second dimension, with the midi-electrophoresis apparatus (GE Healthcare, Uppsala, Sweden, Model: SE-600 Ruby) at 100 V with 60 mA/gel until tracking dye reached lower end of gel.

14.6.3 Staining of gels

a. **Coomassie blue stain**

Staining solution: 0.2% Coomassie brilliant blue G250 stain, 40% methanol, 10% acetic acid

Destaining solution: 40% methanol, 10% acetic acid

After complete gel running, gel was carefully removed and stained with coomassie brilliant blue (CBB R250) staining solution for 4-5 hrs. The gels were then destained with destaining solution for sufficient time. After thorough destaining, the gel was washed in distilled water thoroughly and stored in 5% acetic acid.

b. **Silver staining**

- Incubate the gel in fixer solution (40% ethanol, 10% acetic acid) for 1 hr.

- Wash the gel in water for 30 min.

- Sensitize gels in 0.02% sodium thiosulphate (0.02 g $Na_2S_2O_3$ in 100 mL D. water).
- Wash gel with D. water (3 x 20 sec).
- Place the gel in new staining tray.
- Wash the gel in D. water for 1 min.
- Develop the gel in 3% sodium bi-carbonate and 0.05% formaldehyde.
- Wash the gel with D. water for 20 sec.
- Terminate staining with 5% acetic acid for 5 min.
- Store the gel in 1% acetic acid.

14.6.4 Protein imaging and analysis

The SDS-PAGE/2-DE images were scanned by using Image Scanner III, labscan 6.0 software (GE Healthcare, Bio-Sciences) and analyzed by using ImageMaster2D Platinum 7.0 software (GE Healthcare, Upplasa, Sweeden). Image Master offered comprehensive visualization, exploration and analysis of 2D gel data. The spot detection carried at saliency 150, smoothness 5 and protein minimum area 20. Comparative analysis between pure raw and cooked sheep and buffalo protein gels for spot analysis and followed class analysis by creating different "match sets" were done. After final scanning, images were saved in both Maya Embedded Language (Mel) and Tag Image File Format (TIFF) format. The gel analysis tables, histograms and 3D images created by the software were used for detailed analysis.

14.6.5 In-gel digestion of proteins (Shevchenko *et al.,* 2006)

Reagents required: 25 mM Ammonium Bicarbonate (ABC), 50mM ABC, 10mM DTT, 50mM iodoacetamide (IAA), 0.1% Trifluoroacetic Acid (TFA), Acetonitrile (ACN).

Solution A: 2:1 (ACN: 50mM ABC)

Extraction buffer: 50:50 ACN: H_2O, 0.1% TFA

The desired protein spots were cut into small cubes (1 mm^3) and taken in 1.5mL Eppendorf tube. Rinse the gel with 50μl of 25mM ABC for destaining the gel. Dehydrate and rehydrate the gel with 50μl of solution A for 5 minutes and flushed with 25mM ABC, repeat for 2-3 times. The dehydrated gel was reduced by 50μl of 10mM DTT at 60°C for 1 hour, washed with 50μl of 25 mM ABC followed by alkylation with 50μl of 50mM IAA at room temperature in dark for 30minutes. Dehydrate with solution A for 5 minutes followed by vacuum drying for 5-15 minutes. The gels were subjected to trypsin digestion containing 13μg/μl of sequencing grade trypsin with enough amount of 25mM ABC added to cover the gel pieces and the tubes were incubated at 37°C overnight (12-16hrs). The peptides were extracted by 100μl of extraction buffer and the step was repeated thrice. The collected peptide solution were pooled in a single tube was dried by speed vacuum and mixed with 0.1% TFA for mass spectrometry analysis.

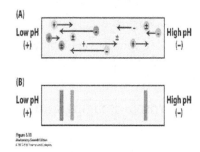

Fig. 14.6. Schematic diagram depicting the principle of Isoelectric focusing (IEF-1st dimension)

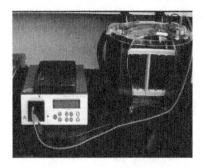
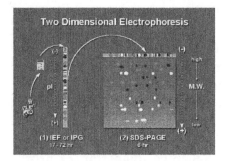

Fig. 14.7. Schematic diagram depicting the principle of SDS-PAGE (2nd dimension)

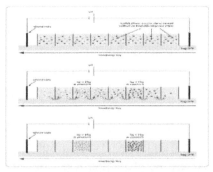

Fig. 14.8. Schematic diagram depicting the principle of OFFGEL electrophoresis

Fig. 14.9. Gel scanner and image analysis

14.7 MASS SPECTROMETRY

Peptide fragment solution were spotted on the MALDI target plate with α0-Cyano-4-hydroxycinnamic acid (CHCA) matrix (5 mg/mL CHCA in 50% ACN/ 0.1% TFA) and allowed to co-crystallize. The protein identification was carried out by 4800 MALDI-TOF/TOF mass spectrometer (AB Sciex, Framingham, MA) linked to 4000 series explorer software (v.3.5.3). Before processing the samples, the instrument was calibrated automatically with peptide standard kit (AB Sciex) having six peptides that include des-Arg1-bradykinin (m/z 904.468), Angiotensin I (m/z 1296.685), Glu1-fibrinopeptide B (m/z 1570.677), ACTH (18–39, m/z 2465.199), ACTH (1–17, m/z 2903.087), and ACTH (7–38, m/z 3657.923). All mass spectra were recorded in a reflector mode within a mass range from 800 to 4000 Da, using a Nd:YAG 355 nm laser. The acceleration voltage used was 20 kV and extraction voltage was 18 kV. All the MS spectra were obtained from accumulation of 900 shots. MS/MS spectra were acquired for the 15 most abundant precursor ions, with a total accumulation of 1000-1500 laser shots and collision energy of 1 kV. The combined MS and MS/MS peak lists were searched using the GPS™ Explorer software version 3.6 (AB Sciex). The search engine used was MASCOT version 2.1 (http://www.martixscience.com) against the Swiss-Prot and NCBI database.

References

Chen, Y.Y., Lin, S.Y., Yeh, Y.Y., Asiao, A.A. and Wu, C.Y. (2005). A modified Protein Procedure for efficient removal of albumin from serum. Electrophoresis, 26: 2117-2127.

Laemmli, U.K. (1970). Cleavage of Structural Proteins during the assembly of the head of bacteriophage T4. Nature, 227, 680-685.

Montowska, M., & Pospiech, E. (2012). Myosin light chain isoforms retain their species-specific electrophoretic mobility after processing, which enables differentiation between six species: 2DE analysis of minced meat and meat products made from beef, pork and poultry. *Proteomics, 12,* 2879–2889.

Naveena, B.M., Deepak, S.J., Jagadeesh Babu, A., Madhava Rao, T., Veeranna, K., Vaithiyanathan, S., Kulkarni, V.V. and Rapole, S. (2017). OFFGEL electrophoresis and tandem mass spectrometry approach compared with DNA-based PCR method for authentication of meat species from raw and cooked ground meat mixture containing cattle meat, water buffalo meat and sheep meat. *Food Chemistry,* 233: 311-320.

Sentandreu, M. A., Fraser, P. D., Halket, J., Patel, R., & Bramley, P. M. (2010). A proteomic-based approach for detection of chicken in meat mixes. *Journal of Proteome Research, 9,* 3374"83.

CHAPTER - 15
■ ■ ■

TECHNIQUES IN MUSCLE CELL CULTURE

Cultured meat is the meat produced by *in vitro* cultivation of animal cells rather than slaughtering of the meat animals. This technology allows meat to be cultured from cells in a fermentor or a bioreactor rather than harvested from livestock on a farm. It is produced by taking a biopsy of the animal cell, and growing it in a controlled, sterile environment outside the animal. Cultured meat is also commonly referred to as, 'Clean meat', '*In vitro* meat', 'Lab grown meat', 'Cell based meat' etc. Cultured meat production is believed to require less water, man power and land area for production of meat as compared to conventional raising of meat animals and slaughter. Cultured meat is also believed to be environment friendly and sustainable technology in view of decreasing water table and shrinking grazing land. Research on production of cultured or lab grown meat is intensifying across the world. Basic techniques required in undertaking research in the area of cultured meat are given in this chapter.

15.1 ISOLATION AND CULTURING OF MYOBLASTS FROM SMALL RUMINANTS

General Equipment

1. Standard humidified tissue culture incubator
2. Biosafety cabinet (Class-II)
3. Phase contrast microscope
4. Cooling centrifuge
5. Autoclave
6. Hot air oven
7. Vortex

Plastic and glassware (Sterile)

1. Petri plates
2. T25 flasks
3. Falcon tubes
4. Serological pipettes
5. Surgical blade
6. Media bottles
7. Gloves and masks
8. Cell strainers
9. Membrane filters and syringe filters

Media and cell culture reagents (Sterile)

1. Dulbeccos Modified Eagle Medium (DMEM) with high glucose
2. Phosphate Buffered Saline (PBS)
3. Fetal Bovine Serum (FBS)
4. Horse Serum (HS)
5. Pencillin & Streptomycin (Penstrep)
6. Trypsin-EDTA
7. Collagen-I
8. Collagenase-I
9. Liquid nitrogen

Steps involved in the culture of myoblast and formation of myotubes

- Collect fresh tissue under sterile condition from the major muscles of the small ruminants
- Wash the collected tissue with PBS containing antibiotic (1% penstrep)
- Suspend the tissue in DMEM with antibiotic (1% penstrep)
- Add 0.1% collagenase I and incubate at 37°C for 50 min and digest the tissue with 0.25% trypsin for 10-20 min (Note: Incubation period should not exceed 30 min as trypsin is toxic to the cells)
- Vortex or do repeated pipetting every 15 min during the incubation time
- The suspension was filtered through a 70/100 μm cell strainer to separate the debris

- Centrifuge the filtrate thrice for 20 min each at 350g and suspend the pellet in growth medium after each centrifugation. (Growth Medium (GM):DMEM+10%FBS+1% Penstrep)
- Pour the suspension onto the culture plate and allow 1 hr incubation at 37°C in 5% CO_2 and 95% humidity (Pre-plating to remove the fastly adhering fibroblasts from slowly adhering myoblasts)
- Collect the supernatant and centrifuge (350g for 3-5 min)
- Suspend the pellet in the growth medium
- Pour the media onto culture plates or flask and incubate at 37°C in 5% CO_2 and 95% humidity
- Change growth medium every 48 hrs
- Repeat the pre-plating method on day 6, 8 and 10 to enrich the culture with myoblasts.
- To differentiate into myotubes plate the myoblast cells onto collagen-I coated 6-well plate in growth medium at 37°C in 5% CO_2 and 95% humidity
- Reaching 80% confluence replace growth medium with DMEM+2% Horse Serum (HS) + 1% Penstrep and culture them at 37°C in 5% CO_2 and 95% humidity for 6 days for the formation of myotubes

   (Sadkowski *et al.*, 2018; Wu *et al.*, 2012)

15.2 ISOLATION AND CULTURING OF MYOBLASTS FROM CHICKEN EMBRYO

Materials required

1. Motor & pestle
2. Petri dishes
3. Curved scissors
4. Rat tooth forceps
5. 15 ml tubes and 50 ml centrifuge tubes
6. T25 flask
7. Sterile absorbent cotton
8. Micro pipettes
9. Aerosol-barrier tips
10. Discard beaker
11. Water bath
12. CO_2 Incubator
13. Refrigerated centrifuge

Chemicals required

1. DMEM/F12 medium
2. Chicken serum
3. 1X PBS
4. 0.25% Trypsin EDTA
5. Surgical spirit

Collection of the sample

- In the hood, wipe the egg with surgical spirit using absorbent cotton. Open the egg from the broad end and place the embryo in the petri dish by using rat tooth forceps
- Carefully separate the leg muscle from the other portion of the body and and place in the petri dish
- Collect the muscle tissue and wash with 1X PBS for 2 to 3 times
- The muscle tissue was made into small pieces with the curved scissors
- These small pieces were triturated using motor & pestle.
- Take the minced muscle into 15ml tube and add trypsin into to it (5ml of medium + 100µl of 0.25% trypsin)
- Minced muscle were incubated at 37°C for 20 min in a water bath with intermittent shaking
- The digested muscle tissue is collected by low speed centrifugation for 3 min at 700rpm
- Collect the supernatant containing the released cells into a fresh 15 ml tube and the pellet with debris was thrown away.
- The supernatant is centrifuged for 7 min at 7000 rpm to pellet the cells.
- **Plating of muscle satellite cells:** The washed cell pellet obtained in the previous step is re-suspended in appropriate growth medium (DMEM/F12+ chicken serum) and directly plated onto T25 flasks.
- The flask is kept in an incubator which is maintained at 37°C in the presence of 5% CO_2

(Baquero-Perez *et al.*, 2012)

15.3 DETERMINATION OF MYOBLAST POPULATION DOUBLING TIME

Cell population doubling time is the time lapse between successive cell divisions. It is determined by counting the number of the cells in culture at different time intervals.

Procedure for determining myoblast population doubling time

- Trypsinize the primary culture cells of sheep/goat/chicken at 70-80% confluency and collect the cells into a 15 ml falcon tube
- Count the cells using heamo cytometer/Neubar chamber
- Seed cells in such a way that there should be 3000 cells/cm^2 area
- Count the cells at 24 hr, 48 hr, 72 hr and 96 hr. At each time point put triplicate cultures and calculate the average number of cells
- Repeat the above steps for at least 3 passages
- Plot a graph with time on X-Axis and number of cells on Y-Axis
- Calculate the average of the doubling time for each passage from the graph

15.4 SDS BASED PROTOCOL FOR EXTRACTION OF EXTRA CELLULAR MATRIX (ECM) OF THE MUSCLE TISSUE

Chemicals required

1. Sodium Dodecyl Sulfate (SDS)
2. Phosphate Buffered Saline (PBS)
3. Triton X-100

Collection and processing of the tissue

- Collect 100 gm of fresh muscle tissue of sheep in PBS
- Remove the fat tissue
- Weigh the tissue (80g) and give thorough wash with PBS

Steps involved

- Rinse the tissue briefly with deionized water and then stir in 1% (wt/vol) sodium dodecyl sulfate (SDS) in phosphate buffered saline (PBS) for 4–5 days, until the tissue is decellularized
- Stir the tissue in 1% (vol/vol) Triton X-100 for 30 min for final cell removal
- Finally, decellularize the muscle tissue and stir it overnight in deionized water to ensure removal of detergents
- Give DNAse/RNAse (1mg/ml) treatment to the ECM for12 hours in PBS
- Wash the ECM with PBS
- Store the ECM in PBS at -80 & for further use (Singelyn *et al.*, 2009)

15.5 TRYPSIN BASED EXTRA CELLULAR MATRIX (ECM) EXTRACTION FROM MUSCLE

Chemicals required

1. PBS
2. Trypsin
3. DMEM+10% FBS
4. Triton X-100

Steps involved

- Rinse the muscle pieces with 1x PBS for 2h
- Incubate the muscle pieces in 0.05% trypsin for 20 min and then inhibit the trypsin by DMEM+10% FBS for 1h
- Wash the pieces in 1X PBS/1% Triton X-100 for 30 min and then with 1x PBS for 1h
- Give DNAse/RNAse (1mg/ml) treatment to the ECM for12 hours in PBS
- Wash the ECM with PBS
- Store the ECM in PBS at -80& for further use

  (Chaturvedi *et al.*, 2015)

15.6 DIGESTION OF EXTRA CELLULAR MATRIX (ECM)

Chemicals required

1. Pepsin
2. Liquid nitrogen
3. Hydrochloric acid
4. Acetic acid

Steps involved

- Poor the liquid nitrogen into the motor and pestle to cool down
- Put the ECM into the motor and pour liquid nitrogen until the liquid nitrogen evaporates. Repeat this step for 2-3 times
- Crush the ECM with pestle until you get fine powder
- Transfer the powder (10mg) into 1 ml pepsin (1mg/ml in 0.01M Hcl)
- Keep at room temperature for 48 hours with slow vortexing

Evaluation of the quality of ECM: Quality of the ECM can be evaluated by DNA quantification, SDS PAGE analysis, electron microscopic studies and proteomic analysis

References

Baquero-Perez, B., Kuchipudi, S.V., Nelli, R.K. *et al.* A simplified but robust method for the isolation of avian and mammalian muscle satellite cells. BMC Cell Biol 13, 16 (2012). https://doi.org/10.1186/1471-2121-13-16

Chaturvedi, V., Dye, D. E., Kinnear, B. F., Van Kuppevelt, T. H., Grounds, M. D., and Coombe, D. R. (2015). Interactions between skeletal muscle myoblasts and their extracellular matrix revealed by a serum free culture system. *PloS one, 10*(6), e0127675.

Sadkowski, T., Ciecierska, A., Oprz¹dek, J., and Balcerek, E. (2018) Breed-dependent microRNA expression in the primary culture of skeletal muscle cells subjected to myogenic differentiation. *BMC genomics, 19*(1), 109

Singelyn JM, DeQuach JA, Seif-Naraghi SB, Littlefield RB, Schup-Magoffin PJ, *et al.* (2009) Naturally derived myocardial matrix as an injectable scaffold for cardiac tissue engineering. Biomaterials 30: 5409–5416

Wu, H., Ren, Y., Li, S., Wang, W., Yuan, J., Guo, X., and Cang, M. (2012). In vitro culture and induced differentiation of sheep skeletal muscle satellite cells. *Cell biology international, 36*(6), 579-587.

■ ■ ■

CHAPTER-16

DETECTION OF CHEMICAL RESIDUES IN MEAT

16.1 DETERMINATION OF RESIDUES OF TETRACYCLINES BY HIGH PERFORMANCE LIQUID CHROMATOGRAPHY

High Performance Liquid Chromatography (HPLC) is one of the most commonly used analytical techniques to separate a wide variety of chemical mixtures. Chromatographic process can be defined as separation technique involving mass transfer between stationary and mobile phase. HPLC utilizes several nonpolar and mixture of water with some polar organic solvent such as acetonitrile or methanol as mobile phase to separate the components of a mixture. The stationary phase is defined as the immobile packing material in the column and it can be a liquid or a solid phase. Depending on the type of the ligand attached to the surface, the adsorbent column could be normal phase ($-OH$, $-NH_2$), or reversed-phase (C5, C8, C 18 CN, NH_2), and even anion ($CH_2NR_3 + OH^-$), or cation ($R-SO_3-H^+$) exchangers. HPLC instrumentation includes a pump, injector, column, detector and data system. The solvent containing mixture of chemical components is forced to flow through a chromatographic column under a high pressure. While moving through the porous packing beads of column, mixture of molecules tends to interact with the surface adsorption sites and then separates into individual molecules. Eventually, each component elutes from the column as a narrow band (or peak) on the recorder. Detection of the eluting components is important, and this can be either selective or universal, depending upon the detector used. Photo Diode Array UV detector (PAD), refractive index (RI), fluorescence (FLU), electrochemical (EC) are the important detectors commonly employed. The response of the detector to each component is displayed on a chart recorder or computer screen and is known as a chromatogram. The amount of resolution is important, and is dependent upon the extent of interaction between the solute components and the stationary phase. The interaction of the solute with mobile and stationary phases can be manipulated through different choices of both solvents and stationary phases. As a result, HPLC acquires a high degree of versatility not found in other chromatographic systems and it has the ability to easily separate a wide variety of chemical mixtures.

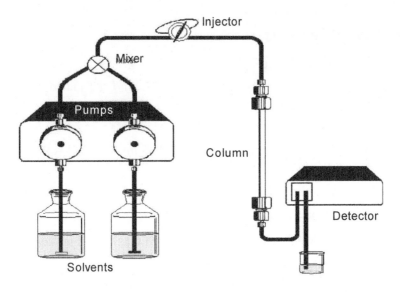

Fig. 16.1. Block diagram of high performance liquid chromatography

Determination of Tetracyclines in foods of animal origin

Instruments & Apparatus: Mortar- pestle or blender, polypropylene centrifuge tubes of 50mL, separating funnels, conical flasks, R.B. flasks, rotary vacuum evaporator, syringe filter (0.2 mm),vials,cartridge C18, Solid Phase Extractor and Centrifuge. Liquid Chromatograph –UV/ Dual Absorbance / MSMS Detector, HPLC Column –reverse phased deactivated silica packing C8 / C18, 5mm, 250 x 4.6 mm id, Turbovap concentrator under Nitrogen.

Chemical/reagents: Citric acid monohydrate AR, Disodium hydrogen phosphate AR, Phosphoric Acid AR, Oxalic acid AR, Acetonitrile LC, Methanol (LC grade), Water (LC grade). **Standards:** USP reference standard of Tetracycline hydrochloride (TC), Oxytetracycline hydrochloride (OTC) & Chlortetracycline hydrochloride (CTC) and their epimers.

Weigh 108±0.1mg each of the Tetracycline hydrochloride in weighing dishes and transfer with methanol into separate 100 mL standard volumetric flask. Make up to volume with Methanol to get 1000-ppm stock solution. Dilute 1 mL of stock to 100 mL with methanol to obtain 10-ppm intermediate solution (I). Again dilute 10 mL of intermediate (I) solution to 100 mL to obtain 1-ppm intermediate solution (II). Pipette out 1 mL, 2 mL, 3 mL, 4 mL & 5 mL of intermediate solution (II) to 10 mL flask and dilute to volume with methanol to obtain appropriate calibration standards.

Preparation of reagents

1. McIlvaine buffer pH 4.0
   a. 0.1M Citric acid monohydrate – 21.01 gm/L (If anhydrous – 19.213 gm/L) in water
   b. 0.2 M Na_2HPO_4 – 28.4 gm/L
      Mix 61.45 mL of solution (a) and 38.55 mL of solution (b). Adjust pH to 4.00 with dilute H_3PO_4
2. 0.01 M methanolic oxalic acid (1.26 gm Oxalic acid /L methanol)

HPLC operating conditions: UV- Detection at 350 nm

1. Mobile Phase – Methanol: Acetonitrile: 0.01 M Oxalic acid in water (73:17:10)
2. Flow rate: 1 mL/minute

HPLC conditions

Column Reverse phase: C8, 2.1 × 150 mm, 5 μm; Flow rate 0.3 mL/min

Mobile phase A: water/0.1% formic acid: B: methanol

Gradient: 0–10 min, B from 5% to 30%, 10–12 min, B from 30% to 40%, 12.5–18 min, B 65%, 18.5–25 min, B 95%, 25.5 min, B 5.0%, Total run 28 min; Post time 5 min

Temperature: 30°C, Injection 5 μL

Test Procedure (Tissue Extraction): Accurately weigh 5.0 ±0.05 gm of well ground homogenized edible portion of muscle tissue into a 50 mL polypropylene centrifuge tube. Macerate/blend the tissue with 20 mL, 20 mL, 10 mL of McIlvaine buffer repeatedly each for at least 30 sec, collecting the extract after each addition. Centrifuge each of the extract at 2500 gm for 10 minutes. Filter the supernatant through GF/B filter paper in a Buchner funnel and moisten with McIlvaine buffer- EDTA solution. Collect the filtrate in a 125 mL sidearm flask, applying gentle vacuum to side arm.

Column Chromatographic separation (SPE)

1. Fit the C 18 cartridge tube on to the SPE extractor.
2. Condition the tube with 20 mL of methanol followed by 20 mL water and discard the eluate.
3. Add the sample filtrate to the tube (maintain a flow rate not exceeding 4 mL/minute)
4. Wash with 20 mL distilled water and let the cartridge dry when the water rinse is complete and continue to draw air through the cartridge for > 2min.
5. Elute with two 5 mL portions of methanolic oxalic acid
6. Collect the elute and evaporate to less than 1 mL under Nitrogen using Water bath at about 40°C 18 m
7. Make up to 1 mL with methanol and filter through 0.2 filter into HPLC vial and inject.

LC Analysis

Inject appropriate volumes (based on response/ Signal to noise ratio and sensitivity of the instrument) of filtered sample extract as well as calibration dilutions of standards for calibration curve into LC system operated isocratically at a mobile phase flow rate of 1.0 mL/minute; with UV detector set at 350 nm and obtain the Chromatogram.

Injection Sequence

1. Inject solvent blank
2. Inject calibration standard(s)
3. Inject the recovery sample
4. Inject the blank sample and verify the absence of analytes
5. Inject sample extract(s).
6. Re-inject the calibration standard at the appropriate level at least after every 20 injections and at the end of the run to verify instrument response.

Calculations: For quantitation of each compound of interest:

1. Review the chromatograms to verify that the analyte peaks are within the retention time windows and that the peaks are integrated correctly.
2. Calculate the normalized peak for each component of interest by dividing the component response by the internal standard response (in case available and used as stated below):

3. Normalized Response Component 1 = Response of Component 1 / Response of Internal Standard c. Generate a linear curve fit to each analyte in standard curve using normalized response to concentration in tissue (μg/kg or ppb).

4. Standard curve must have a correlation coefficient greater than or equal to 0.995.

5. Blank must exhibit a response of less than 5% of the recovery used contemporaneously in the set.

For Confirmation

1. Choose a standard or recovery containing the analyte of interest.

2. Identify 2 product ion peaks in the sample and verify that their peaks are present with a signal to noise ratio > 3. Auxiliary ions may be used if necessary.

3. Identify the retention time of the two product ion peaks in the standard or recovery and in the sample of interest. The sample peak retention times must be within ±5% of the standard or recovery retention times.

4. Calculate the ratio of the response of product ion #2 to product ion #1 in the standardor recovery for the analyte of interest:

5. Ratio = Product ion#2/ Product ion #1 Note: ion ratio should be less than 1. If not,then invert the ratio.

6. Ion ratios determined for each analyte shall be within tolerance limits as described inthe EU document 2002/657/EC incase of positive samples. Suggested tolerances arebased on EU guidelines and range from ± 20% for peaks greater than 50% of the basepeak and to ± 50% for those less than or equal to 10% of the base peak.

Software provided in the instrument can be used for auto quantitation by using linear regression (y=mx+b), where y=peak area/ height, x= Tetracycline concentration in ppb/ μg/kg, m=slope of curve, & b= intercept of y) for samplestaking in to account dilution factor, if any.

Mass Spectroscopy: Conformation of analyzed molecule also important before reporting the results. Sample may contain interfering chemicals that may be misidentified as specific analyte. Data obtained using Mass Spectra can represent the most definitive evidence. The Mass spectra is ideal for coupling with LC owing to its sensitivity and fast response in residue analysis.

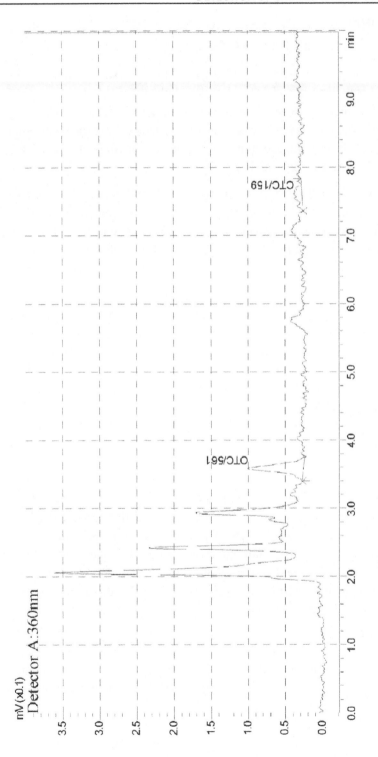

Fig. 16.2. Chromatograph showing OTC and CTC residues in meat samples

16.2 DETERMINATION OF ORGANOCHLORINE PESTICIDE RESIDUES IN MEAT BY GAS LIQUID CHROMATOGRAPHY

The method is employed for estimation of organochlorine pesticides and some of their isomers / metabolites viz., DDT - Dichlorodiphenyl trichlore ethane (p,p'-dichlorodiphenyl dichloroethylene, o , p' – dichlorodiphenyl trichloroethane, p, p' –dichlorodiphenyl trichloroethane, Cyclodiene compounds (aldrin, heptachlor and endosulfan - Endosulfan-α, Endosulfan-β and Endosulfan Sulphate) and HCH-Hexachlorocyclohexane (α, β, γ and δ).

Principle: Fat is extracted from minced meat and partitioned between pertoleum ether and acetonitrile. Further purification is done by chromatography on activated anhydrous sodium sulphate and Florisil column, eluting with mixture of petroleum ether and ethyl ether. Residues in concentrated eluates are identified and quantified by Gas- Chromatography (GC) equipped with Electron capture detector.

Instruments & Apparatus: Soxhlet apparatus, Chromatographic tubes With Teflon stopcocks, Separating funnels, rotary vacuum evaporator/ nitrogen evaporator, Gas Chromatograph equipped with Electron capture detector. For good separation of chlorinated pesticides, a non-polar/fused silicated quartz capillary column and similar are recommended. DBI, capillary column; 30 m long, 0.25 mm id, 0.25 μ film thickness, hamilton syringe - 5/10μl.

Chemical/reagents

1. **Standard of organochlorine pesticide solutions** (A) **Stock solution** 1 mg/mL. Weigh 5 or 10 mg of each pesticide standard reference and transfer into individual corresponding 5 or 10 mL volumetric flask, and accordingly dissolve and make up to 5 or 10 mL with n-Hexane/ nHeptane in the volumetric flask (B) **Intermediate solution** 10 μg/mL. Pipet 1 mL of each stock solution into individual 100 mL volumetric flask. Dilute to volume with hexane. (C) **Working solution** 0.1 μg /mL. Dilute 1 mL of solution B to 100 mL volumetric flask and make the volume with n-hexane.

2. **Other chemicals:** Acetonitrile, Petroleum ether, Ethyl ether, Acetone, Florisil (60/100 PR grade activated at 675°C), Eluting solvent (6% solvent prepared by diluting 60 ml diethyl ether solvent to 1L with petroleum ether), Acetonitrile Saturated with Petroleum ether, Sodium sulphate, Sodium chloride and Sulphuric acid

Cleansing of Glassware

The glassware used were initially washed with tap water and cleansed thoroughly by soaking overnight in detergent solution. The following day, they were again washed in water free of soap and rinsed finally with triple glass-distilled water and dried. The glassware were then rinsed in acetone and dried. All the glassware thus dried were rinsed again with the respective solvents before use.

PREPARATION OF SAMPLE

Extraction of fat

Each sample of meat/byproducts/meat products cut into small pieces and minced separately in a blender. A 15 g portion from each of the minced samples is ground along with twice the quantity of the anhydrous sodium sulphate in a pestle and mortar. Resultant free flowing granular tissue materials are taken into separate thimbles and the extraction of fat is carried out for 6 h in a soxhlet apparatus using petroleum ether at 50 to 55°C as outlined by Tonkabony *et al.* (1981). A few milliliters of petroleum ether is added to the extracted fat and transferred to screw-capped glass vials.

Clean-Up of Samples

The fat extracted from the samples was subjected to clean-up employing the procedures of AOAC (1995) with slight modifications.

Acetonitrile partitioning: The extracted fat is transferred to a 125 ml separator and made up to 15 ml with petroleum ether. Thirty milliliter of acetonitrile saturated with petroleum ether is added and the contents shaken thoroughly with intermittent release of pressure formed during shaking. The layers are allowed to separate and the acetonitrile portion is drained into a one litre separator containing 650 ml water, 40 ml saturated sodium chloride solution and 100 ml petroleum ether. The petroleum ether remained in the 125 ml separator is thrice extracted as above with acetonitrile saturated with petroleum ether at each time. The acetonitrile layer separated at each time is drained into the one litre separator.

The one litre separator containing the acetonitrile extracts is shaken thoroughly. The aqueous layer separated is drained into another one litre separator to which 100 ml petroleum ether is added and shaken thoroughly with care (back extraction into petroleum ether). After discarding the aqueous layer, the petroleum ether portion is combined with that in the first one litre separator, washed with two 100 ml portions water and the washings are discarded.

Sodium sulphate column clean up: The remaining petroleum ether is passed through a 50 x 25 mm od (outer diameter) column of activated anhydrous sodium sulphate and collected into a 500 ml concentrating flask along with two rinses of the column using a few milliliters of petroleum ether and concentrated to less than 10 ml in a vacuum evaporator.

Florisil column clean up: A 22 mm id (inner diameter) x 100 mm florisil column is prepared by sandwiching florisil between two layers of anhydrous sodium sulphate in a chromatographic tube plugged with non-absorbent cotton at the bottom. The column is pre-washed with 50 ml of petroleum ether followed by 20 ml of the eluting solvent and kept ready for clean up of the samples. The petroleum ether concentrate is transferred drop by drop into the florisil column prepared. Elution is carried out with 200 ml of 6% eluting solvent at 40-45 drops per minute. The elute is collected in a 500 ml concentrating flask and dried completely in a vacuum evaporator. Reconstitution based on the concentration of the residues with acetone of known volume is carried out.

QuEChERS method for sample extraction and clean up: (Quick, Easy, Cheap, Effective, Rugged, and Safe) methodology is an effective and versatile alternative technique to traditional pesticide analysis and has been applied for the analysis of various chemical residues including pesticides in livestock products. QuEChERS involves a microscale extraction with acetonitrile combined with dispersive solid-phase extraction (d-SPE) using primary secondary amine (PSA) or other sorbents for purifying the extract. This methodology shows several advantages over traditional pesticide analysis such as quick and single step sample preparation instead of a series of time-consuming solvent extractions, reduced needs for reagents and labware and versatility. This methodology is suitable for analyzing chemical residues in livestock products with various contents of fat, attaining adequate recoveries that could make it a good option for routine sample screening. This method involves placing 10 gram of homogenized ground meat into a 50-mL centrifuge tube and into which 2 mL of water and 10 mL of acetonitrile (ACN) is added and then the tube is shaken vigorously for 1 minute. Then add 4 gram of anhydrous magnesium sulfate and 1 gram of sodium chloride into centrifuge tube and shaken vigorously for 1 minute. Centrifuge for 3 - 5 minutes at 4000 rpm and then transfer 1-mL aliquot of supernatant to a 2-mL dSPE cleanup tube that contains 150 mg of magnesium sulfate, 50 mg PSA sorbent, and 50 mg C_{18} sorbent. Shake vigorously for 1 min. Transfer a portion of the supernatant to the vial for GC analysis.

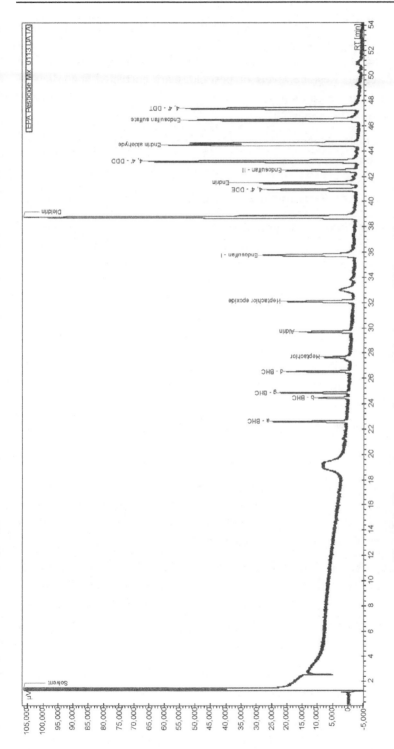

Fig. 16.2. Chromatogram of organochlorine pesticide residues

Recovery Study: Meat samples were fortified with the working standards (0.01 and 0.1 ppm) to estimate the recovery by following the procedure described above and compared the results with the unfortified samples to ascertain the efficiency of extraction. Residue levels of pesticides are corrected according to their recoveries and expressed as mg/kg (ppm). Recoveries of added compounds through method should be about 80 %.

Estimation of Pesticide Residues

GC operating condition:

Temperature:

Oven: Temperature Increment Schedule:

Temperature increment rate	Temperature (°C)	Holding time (Minutes)
	80	2.0
5.0	150	2.0
2.0	180	5.0
3.0	210	5.0
3.0	220	10.0

Injector	:	260°C
Detector	:	300°C
Carrier gas		
Gas		Nitrogen (N$_2$)
Flow rate		5 ml/min
Make-up gas		55 ml/min

Preparation of standards chromatogram: One micro litre of 0.01 and 0.1 ppm of the pesticides mixer was injected into a gas chromatograph equipped with an electron capture detector. This injection mode was splitless. The retention time along with the height and areas of the peak were recorded.

Samples analysis: One micro litre of each sample reconstituted in n-hexane was injected into the Gas Chromatograph as followed for the standards and the peak areas along with the retention time recorded and compared with that of the standards. The concentration of pesticide residues in mg/kg was calculated (IS: 5864-1983 and IS: 6169-1983) on fat basis as follows:

$$\text{Pesticide residues (mg/kg)} = \frac{H_s}{H_{std}} \times \frac{M}{M_1} \times \frac{V}{V_1}$$

Where,

H_s	=	Peak height of the sample
H_{std}	=	Peak height of the standard
M	=	μg of standard injected
M_1	=	Mass, in g of the sample
V	=	Volume of final extract in ml
V_1	=	μl of the sample injected

References

AOAC 1995 Official methods of analysis 17th ed. Association of Official Analytical Chemists. Arlington, VA. No. 970.52.

AOAC Official Methods of Analysis, 2005, 995.09, Ch. 23.1.17, p-22 to 26

Garcia, CV and Gotah A. 2017. Application of QuEChERS for determining xenobiotics in foods of animal origin. Journal of Analytical Methods in Chemistry. doi.org/10.1155/2017/2603067.

Sadkowski, T., Ciecierska, A., Oprz¹dek, J., and Balcerek, E. (2018). Breed-dependent microRNA expression in the primary culture of skeletal muscle cells subjected to myogenic differentiation. *BMC Genomics*, *19*(1), 109.

Tonkabony S E H, Afshar A, Ghazisaidi K, Langaroodi F A, Messchi M and Ahmadi Z 1981 Chlorinated pesticide residues in meat and fat from animals raised in Iran. Journal of American Oil Chemists Society 58: 89-91.

Venugopal, G., Kalpana, S., Baswa Reddy, P and Muthukumar, M. 2019. Quantitative determination of residual tetracyilnes and fluoroquinolones in field samples of fish using RP-HPLC.In compendium of "World brackishwater aquaculture conference 2019" Page 197.

Wu, H., Ren, Y., Li, S., Wang, W., Yuan, J., Guo, X., ... & Cang, M. (2012). In vitro culture and induced differentiation of sheep skeletal muscle satellite cells. *Cell Biology International*, *36*(6), 579-587.

CHAPTER - 17
■ ■ ■
SENSORY EVALUATION OF
MEAT PRODUCTS

Sensory evaluation uses the entire human senses; smell, touch, sight, hearing and taste, to measure the characteristics of a food product in a scientifically controlled manner. Sensory analysis is used to evaluate attributes of products, testing for differences (qualitative and quantitative), or testing for acceptability. The Sensory Evaluation Unit can also use sensory analysis for product development, trouble-shooting and quality control.

It is defined as a scientific discipline used to measure, analyze and interpret reactions to those characteristics of materials as they are perceived by the senses of sight, smell, taste, touch and hearing.

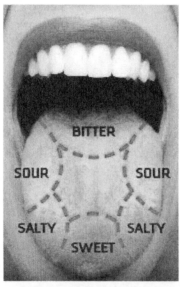

Fig. 17.1. Picture depicting location of different tast buds on tongue

The objectives of sensory evaluation are

● To study the sensory evaluation of meat in its use for research

● To demonstrate the methods of meat cookery used for sensory evaluation.

● To describe the correct procedures for temperature control, coring, shearing and serving samples to panels.

The evaluation of meat products can be divided to into two categories

● Taste panel evaluation

● Measurement of sensory attributes by help of machines like Instron tester and Warner – Bratzler shear force.

In the taste panel evaluation sensory analysis is used to establish differences and to characterise and measure sensory attributes of products or, to establish whether product differences are acceptable or un-acceptable and noticeable to the consumer.

Sensory Evaluation is used in:

● Shelf life studies of a product

● Product matching to compare and modify the sensory characteristics of one product to be in line with sensory characteristics of another similar product.

● Product mapping, where sensory profiles are produced for a range of products in the market place, helps to identify product position in respect to competitive products, and to identify gaps in the product range which may be successfully filled by new product development.

● Product specification and quality control. This is useful in normal commercial practice.

● Product reformulation in order to cope with the new legislation, additives or ingredients available.

● Taint potential. Sensory evaluation can establish whether a taint problem is likely to develop due to painted surface, use of disinfectant, packaging material or local atmosphere, etc.

● Over all product acceptability.

Table 17.1: Types of the Tests Used in Sensory Evaluation

Test	No. of assesses required
Sensory test	
Difference test	
Paired test	A-30, SA-20

Test	No. of assesses required

This test is used to determine if two samples differ in a specific character. It is a directional test with a named attribute, e.g. the assessor is presented with two samples and asked which sample is tougher or more juicy.

| Triangular test | A-24, SA-18 |

This test is used to determine an unspecified sensory difference between two treatments. The assessor is presented with three samples, advised that one may be different, and asked to identify which is the different sample.

| Two out of five test | A-312, SA-32 |

Multiple sample tests can be used to determine the differences between two treatments. In this test, the number of presentations is increased to five and the assessor is instructed to identify the sample, which are similar.

| Duo-two test | A-32, SA-20 |

This test is used to measure unspecified difference between samples. The assessor is presented with one sample and then a pair of samples and asked to identify which of the pair matches the first sample.

| Ranking test | A-30 |

A ranking test with a specified attribute is used to establish a magnitude of differences between samples. The assessors are presented with three or more coded samples and asked to rank them in order according to a single specific attribute e.g. mouth coating in sausage or harsh taste.

Rating test	A-20, SA-8
Descriptive test	SA –8
Acceptance test	
Two sample preference test	A-50
Multisampling ranking for preferences	A-50

This is a form of ranking test except that the specified attribute is preference. Analysis is also known by the Friedman rank test.

| Hedonic ranking | A-70 |

In this test the assessor is asked to indicate the extent of liking for the product from extreme dislike to extreme like. A popular scale is the nine point hedonic scale (Peryam and Pilgrim, 1957). In this each descriptor

Test	No. of assesses required
is assigned a value and it is usually assumed that it is an equal interval scale. An alternative approach is to score on a continuous line scale with the extremes at either end. The distance of the mark along the line is then used as rating (Table 9.2).	
Magnitude ranking	A-70

A= assessor = a person carrying out the sensory test

SA- selected assessor = assessors who have been selected because of their sensitivity and ability to perform the test in question.

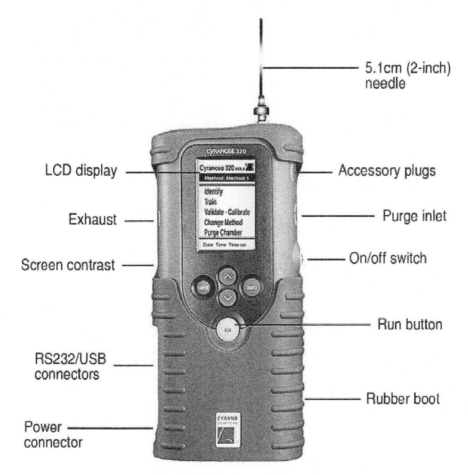

Fig. 17.2. Cyranose 320 e-nose

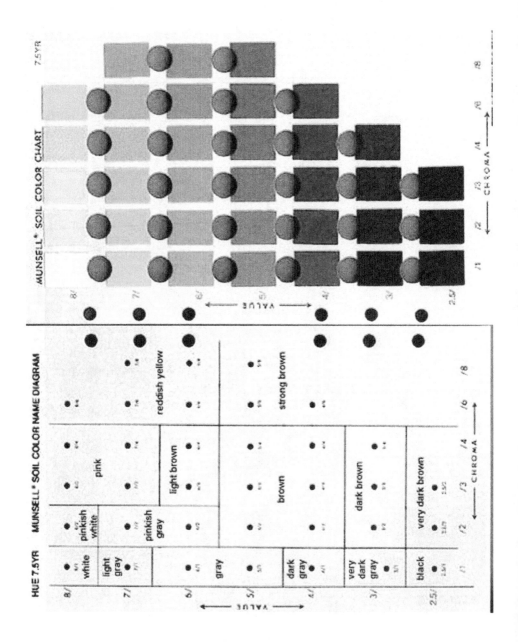

Table 17.2: Nine Point Hedonic Rating Performa

Sensory evaluation of _____

Name _____   Date _____

Treatment

Attributes	9	8	7	6	5	4	3	2	1
Appearance	Very Excellent	Excellent	Very good	Good	Fair	Slightly poor	Moderately	Very poor	Extremely Poor
Flavour Texture	Very Extremely desirable	Extremely desirable	Very Desirable	Moderately Desirable	Slightly Desirable	Slightly Undesirable	Moderately undesirable	Very Desirable	Extremely desirable
Juiciness	Very Extremely juicy	Extremely juicy	Very juicy	Moderately Juicy	Slightly juicy	Slightly dry	Moderately Dry	Very dry	Extremely Dry
Mouth coating	None	None	Practically None	Traces	Slight	Moderately	Slightly Abundant	Moderately Abundant	Extremely Abundant
Overall Acceptability	Very Extremely acceptable	Extremely acceptable	Very acceptable	Moderately acceptable	Slightly acceptable	Slightly unacceptable	Moderately unacceptable	Very Unacceptable	Extremely unacceptable
Sample		Appearance	Juiciness	Texture	Flavour	Mouthcoating	Overall acceptability		

Remarks

Signature

References

David, H Lyon; Mariko-A. Francombe; Terry, A Hasdell and Ken Lawson (1994) In Guidelines for sensory analysis in food product development and quality control. Published by Chanpman and Hall , 2-6. Boundary Row, London, SE 1 8 HN, UK.

Peryam, D.R. and Pilgrim, F.J. (1957) hedonic scale method for measuring food preferences, Food Technology, 11 (9) 9-14.

■ ■ ■

TECHNIQUES FOR CHARACTERIZATION OF SLAUGHTERHOUSE EFFLUENT

18.1 DETERMINATION OF CHEMICAL OXYGEN DEMAND IN SLAUGHTERHOUSE EFFLUENT

The quantum of organic matter in water is very critical and waste that's high in organic matter need to be treated to reduce the amount of organic waste before discharging into receiving waters. Discharging of water with high organic content into natural waters led to consumption of the oxygen present in the receiving water by microbes as part of the breakdown of organic waste. This oxygen depletion along with nutrient rich conditions is called eutrophication, a condition of natural water that can lead to the death of animal life.

Principle

Chemical oxygen demand (COD) is an indirect measurement of the amount of organic matter in a sample. The method determines the amount of oxygen required for chemical oxidation of organic matter using a strong chemical oxidant, such as, potassium dichromate. This test is widely used to determine degree of pollution in water bodies and their self-purification capacity, efficiency of treatment plants, pollution loads, and provides rough idea of Biochemical oxygen demand (BOD) which can be used to determine sample volume for BOD estimation. Most of the organic matters are destroyed when boiled with a mixture of potassium dichromate and sulphuric acid producing carbon dioxide and water. A sample is refluxed with a known amount of potassium dichromate in sulphuric acid medium and the excess of dichromate is titrated against ferrous ammonium sulphate. The amount of dichromate consumed is proportional to the oxygen required to oxidize the oxidizable organic matter. Potassium dichromate is a hexavalent chromium salt that is bright orange in color. Once dichromate oxidizes a substance, it's converted to a trivalent form of chromium, which is a dull green color. Digestion is performed on the samples with a set amount of the oxidant, sulfuric acid, and heat (150°C). Metal salts are usually included to suppress any interferences and to catalyze the digestion. The digestion typically takes two hours to perform. During the digestion, it's necessary to have excess oxidant; this ensures complete oxidation of the sample. As a result, it's important to determine the quantity of excess oxidant. The two most common methods for this are titration and colorimetry.

Reagents

Standard Potassium dichromate ($K_2Cr_2O_7$) digestion solution, 0.01667M: Add to about 500 mL distilled water 4.903 g $K_2Cr_2O_7$, primary standard grade, previously dried at 150°C for 2 h, 167 mL conc. H_2SO_4, and 33.3 g $HgSO_4$. Dissolve, cool to room temperature, and dilute to 1000 mL.

Sulfuric acid reagent: Add H_2SO_4 at the rate of 5.5 g Ag2SO4/kg H_2SO_4 or 10.12 g silver sulphate/L H_2SO_4. Let stand 1 to 2 d to dissolve and mix. This accelerates the oxidation of straight chain aliphatic and aromatic compounds.

(1 Kg = 543.47826 mL of H_2 SO_4 and take 20.24 g of Ag_2SO_4 to 2 L of H_2 SO_4 or 22.264 g of Ag_2SO_4 to 2.2 L of H_2 SO_4)

Ferroin Indicator solution: This indicator is used to indicate change in oxidation-reduction potential of the solution and indicates the condition when all dichromate has been reduced by ferrous ion. It gives a very sharp brown color change which can be seen in spite of blue color generated by the Cr^{3+} ions formed on reduction of the dichromate.

Standard ferrous ammonium sulfate titrant (FAS), approximately 0.10M: Dissolve 39.2 g Fe $(NH_4)_2(SO_4)_2.6H_2O$ in distilled water. Add 20 mL conc. H_2SO_4, cool, and dilute to 1000 mL. Standardize solution daily against standard $K_2Cr_2O_7$ digestion solution as follows: Pipette 5.00 mL digestion solution into a small beaker. Add 10 mL reagent water to substitute for sample. Cool to room temperature. Add 1 to 2 drops diluted Ferroin indicator and titrate with FAS titrant.

$$\text{Molarity of FAS solution} = [V_{K2Cr2O7} \times 0.1] / (V_{FAS})$$

Where: $V_{K2Cr2O7}$ = volume of $K_2Cr_2O_7$ (mL); V_{FAS} = volume of FAS (mL)

Procedure

1. Collect approximately 1-L, or a minimum of 100 ml, of a representative sample in a plastic or glass bottle following conventional sampling techniques.

2. Refrigerate samples at 0 to 4°C from the time of collection until the time of analysis.

3. Analyze the sample within 28 days of collection.

4. Blend samples containing settleable solids in a blender for two minutes. Homogenizing the sample in this way ensures the solids are evenly distributed throughout the sample, thereby increasing accuracy and reproducibility of results.

Titrimetric Method of COD

In the titration method for determining COD, the excess dichromate is reacted with a reducing agent, ferrous ammonium sulfate. As the ferrous ammonium sulfate (FAS) is added slowly, the excess dichromate is converted into its trivalent form. As soon as all the excess dichromate reacts, an equivalence point is reached. This point means that the amount of ferrous ammonium sulfate added is equal to the amount of excess dichromate. Color indicators can also signal this endpoint.

1. Wash culture tubes and caps with 20% H_2SO_4 before using to prevent contamination.

2. Place sample (2.5mL) in culture tube and Add $K_2Cr_2O_7$ digestion solution (1.5 mL).

3. Carefully run sulphuric acid reagent (3.5 mL) down inside of vessel so an acid layer is formed under the sample-digestion solution layer and tightly cap tubes or seal ampules, and invert each several times to mix completely.

4. Place tubes in block digester preheated to 150°C and reflux for 2 h

5. Cool to room temperature and place vessels in test tube rack. Some mercuric sulfate may precipitate out but this will not affect the analysis.

6. Add 1 to 2 drops of Ferroin indicator and stir rapidly on magnetic stirrer while titrating with standardized 0.10 M FAS.

7. The end point is a sharp color change from blue-green to reddish brown, although the blue green may reappear within minutes.

8. In the same manner reflux and titrate a blank containing the reagents and a volume of distilled water equal to that of the sample.

9. COD is given by: COD (mg O_2 /L) = [(A-B) × M ×8000) / (V_{sample})
    Where A- voume of FAS used for blank (ml); B- volume of FAS used for samples(ml) and M- molarity of FAS; 8000 = milli equivalent weight of oxygen (8) ×1000 mL/L.

Colorimetric Method of COD

The consumption of dichromate could be assessed by measuring the change in the absorbance of the sample. The samples absorb at particular wavelengths due to the color of trivalent chromium (Cr^{3+}) and hexavalent chromium (Cr^{6+}). The amount of trivalent chromium in a sample after digestion is quantified by measuring the absorbance of the sample at a wavelength of 600 nm in a photometer or spectrophotometer. Alternatively, the absorbance of hexavalent chromium at 420 nm can be used to determine the amount of excess chromium at the end of digestion to determine COD values.

1. Digest your samples and a reagent blank (The reagent blank is just a sample of deionized water that's treated the same as the actual samples)
2. Let the digested samples and blank cool
3. Zero the instrument using the blank vial and read the samples

    (AWWA, WEF, APHA, 1998; Sawyer, 2000)

18.2 DETERMINATION OF BIOCHEMICAL OXYGEN DEMAND (BOD) IN SLAUGHTERHOUSE EFFLUENT

Objective: The biochemical oxygen demand (BOD) test measures the strength of the wastewater/industrial effluent/surface water by estimating the quantum of oxygen consumed by the aerobic organisms as they stabilize the organic matter under controlled conditions of specific duration and temperature.

Principle

The biochemical oxygen demand (BOD) test is based on the principle that in the presence of adequate amount of oxygen, biological decomposition of organic matter by aerobic microbes will progress until all waste is consumed. As this test is based on the exact quantification of dissolved oxygen at the start and end of a 5 days duration in which the waste water sample is kept in dark, controlled incubated conditions at 20°C, it is also named as "BOD5". The change in the concentration of dissolved oxygen over five days duration indicates the "oxygen requirement" for respiration by the aerobic microbes in the water sample.

Manometric Respirometric Test:

The manometric method for BOD determination is based on the fact that the organic matters, which are consumed by microbes in the presence of O_2 in waste water, are biochemically oxidised and broken down into to carbon dioxide and inorganic salts (mineralisation). Then the oxygen which is converted to carbon dioxide is absorbed from the gas phase of the sample using potassium hydroxide (Hütter, 1984). The drop in pressure due to removal of carbon dioxide is proportional to the quantity of O_2 utilized by microbes.

The manometric method based BOD measurement system consisting of sample bottle and the BOD sensor, which indicates a closed system. In the bottle, above the sample itself, is a specified quantum of air. During the BOD measurement, the microbes present in wastewater samples consume dissolved oxygen in the sample and release carbon dioxide. At the same time, the released carbon dioxide will be absorbed by the potassium hydroxide kept above the sample in the seal gasket. This generates a decrease in pressure within the system. From the decrease of pressure, the sensor electronics calculates a value that is in direct proportion to the BOD value in mg oxygen / litre. To achieve consistent and reproducible results, the estimation needs to conducted at a constant temperature of 20°C.

Pre-treatment of the sample

1. The pH value of the effluent sample needs to be optimized before analysis. The optimum pH value for biochemical oxidation ranges between pH 6.5 and 7.5. If the pH value of the sample is higher or lower, it should be pre-adjusted. Any significant deviation in pH value will lead to a lower BOD value. Higher pH value could be reduced by addition of dilute hydrochloric acid (1 mol/l) or dilute sulphuric acid (1 mol/l). Similarly, lower pH value could be adjusted with a sodium hydroxide solution (1 mol/l).

2. Sample must be free from chlorine. If sample contains chlorine, addition of dechlorination chemical like sodium sulfite prior to analysis is necessary.

3. Presence of existing adequate microbiological population in the water sample is very essential. In case, the water sample has inadequate or unknown microbial population, addition of "seed" solution of bacteria along with essential nutrients is needed.

4. The water sample needs to be homogenized and mixed well and allowed to settle for a short while.

5. Accurate quantity of representative portion of sample need to be taken in the sample bottle. It is advisable that each sample need to be tested two or three times.

BOD measurement procedure

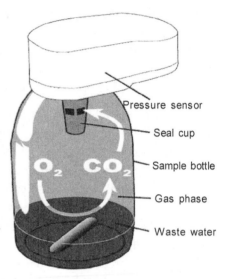

Pressure sensor

Seal cup

Sample bottle

Gas phase

Waste water

1. Set the temperature of the thermostatically controlled cabinet to 20±0.1°C and switch on. Then keep the stirring system into the cabinet.

2. Take exact quantity of the sample and then add into test bottle. The range of the measurement system should be selected to ensure that the expected readings will be roughly within the upper half of the scale (Table 19.1). For example, where BOD values of 250 mg/l are expected, the range 0-400 mg/l would be ideal. Similarly the nitrification inhibitor solution (Allyl Thiourea) needs to be added as per BOD measurement. This is very pertinent for the low range 0 - 40 mg/l (for example, when checking discharges from effluent treatment plants).

3. Add a clean magnetic stir bar to each test bottle and fill the dry seal cup with 3-4 drops of potassium dichromate solution (for binding of the carbon dioxide) and place the seal cup in the test bottle.

4. Place the BOD sensor head on the test bottles and carefully screw them in such a way that system is absolutely tight. Then place the BOD bottle on its bottle rack.

Table 19.1. BOD measurement ranges with the associated sample volumes and the required quantity of nitrification inhibitor (ATH)

BOD measurement range (mg/l)	Sample volume (ml)	Nitrification inhibitor solution (drops)
0 - 40	428	10
0 - 80	360	10
0 - 200	244	5
0 - 400	157	5
0 - 800	94	3
0 - 2000	56	3
0 - 4000	21,7	1

5. Incubate the sample for 5 days at 20°C

6. Measurement may be carried out with the selected sensor every hour/2 hrs upto 1-2 days and thereafter once in every 24 hrs.

7. The measured value at the end of 5 days test period should be within the tolerance range.

(APHA, 1992; Hutter, 1994)

References

APHA. 1992. Standard methods for the examination of water and wastewater. 18th ed. American Public Health Association, Washington, DC.

AWWA, WEF, APHA, 1998, Standard methods for the examination of water and wastewater (Methods: 5220 C. Closed Reflux Titrimetric Method)

Hutter 1994. Wasser und Wasseruntersuchung, 6. Auflage, Otto Salle Verlag Frankfurt am Main.

Sawyer, C.N., McCarty, P.L., and Parkin, G.F. 2000. *Chemistry for Environmental Engineering* 4th Edition. Tata McGraw-Hill Publishing Company Limited.

C H A P T E R - 19

■ ■ ■

STORAGE AND HANDLING OF LABORATORY CONSMABLES

Storage and handling of laboratory consumables

Safely storing chemicals in a laboratory requires diligence and careful consideration. Correct use of containers and common lab equipment is critical.

- All chemicals should be labeled with appropriate hazard warnings and dated upon receipt in the lab and on opening

- Provide a specific storage space for each chemical, and ensure return after each use.

- Store volatile toxics and odoriferous chemicals in ventilated cabinets.

- Store flammable liquids in approved flammable liquid storage cabinets. Small amounts of flammable liquids may be stored in the open room.

- Separate all chemicals, especially liquids, according to compatible groups. Follow all precautions regarding storage of incompatible materials. Post a chemical compatibility chart for reference, both in the lab and next to chemical storage rooms.

- Use appropriate resistant secondary containers for corrosive materials. This protects the cabinets and will catch any leaks or spills due to breakage.

- Seal containers tightly to prevent the escape of vapors.

- Use designated refrigerators for storing chemicals. Label these refrigerators CHEMICAL STORAGE ONLY—NO FOOD. Never store flammable liquids in a refrigerator unless it is specifically designed and approved for such storage. Use only explosion-proof (spark-free) refrigerators for storing flammables.

- Do not store large, heavy containers or liquids on high shelves or in high cabinets. Instead store these at shoulder level or below. Do not overcrowd shelves. Each shelf should have an anti-roll lip.

- Store highly toxic or controlled materials in a locked, dedicated poison cabinet.

- Avoid storing bottles on the floor unless they are in some type of secondary containment.

- Do not store chemicals near heat sources or in direct sunlight.

- Chemicals should not be stored in fume hoods. Excessive containers interfere with air flow and hood performance. Only chemicals in actual use should be in the hood.

- Do not store anything on top of cabinets. Ensure at least 18 inches of clearance around all sprinkler heads to avoid interference with the fire suppression system.

- Avoid using bench tops for storage. These work spaces should contain only chemicals currently in use.

- Do not store chemicals indefinitely. Humidity causes powders to cake or harden. Liquid chemicals evaporate. Ensure all manufacturers' expiration dates are strictly followed. Pay special attention to reactive or dangerous compounds. Dispose of all outdated, hardened, evaporated, or degraded materials promptly.

- First aid supplies, emergency phone numbers, eyewash and emergency shower equipment, fire extinguishers, spill cleanup supplies and personal protective equipment should be readily available and personnel trained in their use.

- Strong acids and other corrosive chemicals should be stored in cabinets lined with lead and located near a fume hood and emergency shower. Volatile corrosive chemicals should be protected from heat. When storing and handling dangerous chemicals avoid producing dangerous mixtures.

Flammable Liquid Storage and Handling

- A flammable liquid is any liquid having a flashpoint (minimum temperature at which a liquid gives off enough vapor to form an ignitable mixture with air near the surface of the liquid) at or below 93°C. To prevent fires, hazardous liquids require special precautions in storage, handling and use.

- Storing flammable liquid in safety cans and safety cabinets help reduce the hazards associated with flammables. Safety can are approved container, of not more than 5-gallons capacity, having a spring-closing lid and spout cover and so designed that it will safely relieve internal pressure when subjected to fire exposure. Storage cabinets for storing flammable liquid are designed and constructed to limit the internal temperature to not more than 162.7°C when subjected to a 10-minute fire test and the cabinets must be labeled in conspicuous lettering, "Flammable - Keep Fire Away."

APPENDIX

■ ■ ■

WEIGHTS AND MEASURES

---

1 kg	=	2.205 lb
1 gallon	=	0.035 oz
1 oz	=	28.35 g
1 lb	=	0.454 kg
1mg	=	0.015 grain 1grain = 64.8 mg
1MT	=	1000 kg = 2205 lb
1mg /kg	=	1 ppm
1 mg / 1000 kg	=	1 ppb
1 mg/kg	=	1 ppb
1 gal	=	3.785 lit
1 quart	=	0.946 lit
1 pint	=	473.2 ml
1 oz	=	29.57 ml
1 teaspoon	=	5 ml
1 table spoon	=	15 ml
1 meter	=	39.37 in
1 yard	=	91.44 cm = 3 ft
1 acre	=	43560 sq. ft or 4047 sq. mt. or 0.405 hectare
1 ft	=	30.48 cm
1 in	=	2.54 cm
1 meter	=	1000 mm
1 mm	=	1000 m m
1 mm	=	1000 nm

1 nm	=	1000 ppm
1A^0	=	10^{-10} m or 10^{-7} mm
1 nm	=	10A^0
1 nm	=	10^{-9} m

To convert Mpa to psi multiply by 145

To convert N / cm to lbf / ft divide by 14.6

1 kg / sq. cm = 14.223 psi

1 psi = 0.07 kg / sq. cm

Temperature conversion:- C / 5 = f – 32 / 9, C = Centigrade; f = Fahrenheit **Power consumption** One KW per hour is equivalent to one unit of electrivity consumptionOne HP motor for one hour consumes 0.747 KW, or one KW is equal to 1.34 HP